Sustainability, Well-Being and Socio-Environmental Quality

Edited by
Luca Salvati

ANTHEM PRESS

Anthem Press
An imprint of Wimbledon Publishing Company
www.anthempress.com

This edition first published in UK and USA 2026
by ANTHEM PRESS
75–76 Blackfriars Road, London SE1 8HA, UK
or PO Box 9779, London SW19 7ZG, UK
and
244 Madison Ave #116, New York, NY 10016, USA

British Library Cataloguing-in-Publication Data
A catalogue record for this book is available from the British Library.

Library of Congress Cataloging-in-Publication Data: 2025950399
A catalog record for this book has been requested.

ISBN-13: 978-1-83999-875-1 (Hbk)
ISBN-10: 1-83999-875-X (Hbk)

Cover Credit: Photograph by Luca Salvati

This title is also available as an eBook.

TABLE OF CONTENTS

Notes on Contributors

Vasiliki Benaki, Hellenic Statistical Authority, Piraeus 46 & Eponiton Street, EL-18510, Piraeus, Greece; vasbenaki@statistics.gr

Gianluca Egidi, Department of Agricultural and Forest Sciences, University of Tuscia, Via San Camillo de Lellis, IT-01100 Viterbo, Italy; egidi.gianluca@unitus.it

Luigi Fici, Department of Economics, Engineering, Society and Business Organization, Via S. Maria del Paradiso 47, I-01100 Viterbo, Italy; luigi.fici@unitus.it

Ioannis Konaxis, Department of Tourism Studies, University of Piraeus, Odos Karaoli & Dimitriou 80, EL-18534 Piraeus, Greece; ikonaxis@unipi.gr

Pierpaolo Lombardi, Gruppo Maurizi – Sicurezza alimentare, ambientale e sul lavoro, Via Pellaro 22, IT-00178 Rome, Italy; pierpaolo.lombardi@gruppomaurizi.it

Marco Maialetti, Independent researcher and lawyer, Circonvallazione Clodia 82, IT-00195 Rome, Italy; avv.marcomaialetti@gmail.com

Maurizio Masi, Department of Economics, Engineering, Society and Business Organization, Via S. Maria del Paradiso 47, I-01100 Viterbo, Italy; maurizio.masi@unitus.it

Enrico Maria Mosconi, Department of Economics, Engineering, Society and Business Organization, University of Tuscia, Via del Paradiso, IT-01100 Viterbo, Italy; enrico.mosconi@unitus.it

Bogdana Nosova, Department of Social Communications, Department of Social Communications, Educational and Scientific Institute of Journalism Taras Shevchenko National University of Kyiv, 64/13 Volodymyrska Street, UK-01601 Kyiv, Ukraine; bogdana.nosova@gmail.com

Bünyamin Özkür, Department of Economics, Faculty of Economics and Administrative Sciences, Manisa Celal Bayar University, Faculty of Economics and Administrative Sciences, Sehit Prof. Dr. Ilhan Varank Campus, Muradiye, 45140, Manisa, Turkey; bunyamin.ozkur@cbu.edu.tr

Giovanni Quaranta, Dipartimento di Matematica, Informatica ed Economia, Università degli Studi della Basilicata, Campus di Macchia Romana, Via dell'Ateneo Lucano 10, IT-85100 Potenza, Italy; giovanni.quaranta@unibas.it

Jesus Rodrigo Comino, Department of Regional Geographic Analysis and Physical Geography, Facultad de Filosofía y Letras, Campus Universitario de Cartuja, ES-18071 Granada, Spain; jesusrc@ugr.es

Luca Salvati, Department of Methods and Models for Economics, Territory and Finance (MEMOTEF), Faculty of Economics, Sapienza University of Rome, Via del Castro Laurenziano 9, IT-00161 Rome, Italy; luca.salvati@uniroma1.it

Rosanna Salvia, Dipartimento di Matematica, Informatica ed Economia, Università degli Studi della Basilicata, Campus di Macchia Romana, Via dell'Ateneo Lucano 10, IT-85100 Potenza, Italy; rosanna.salvia@unibas.it

Adele Sateriano, Mediterranean Foundation for Sustainable Development (MEDES), Contrada San Licandro 1, IT-84029 Sicignano degli Alburni (Salerno), Italy; adele@medes.eu

Donato Scarpitta, Independent researcher and architect, Contrada Salise, IT-84067 Santa Marina (Salerno), Italy; scarpitta.donato@libero.it

Chapter 1

IN-BETWEEN SOCIETY AND LANDSCAPE: LOCAL DEVELOPMENT AND THE OPTIMAL MANAGEMENT OF 'SUSTAINABLE' DISTRICTS

Adele Sateriano, Ioannis Konaxis, Maurizio Masi, Luigi Fici, Marco Maialetti, Donato Scarpitta, Bogdana Nosova and Luca Salvati

1. Introduction

While generic, the definition of ecological sustainability is routinely associated with Brundtland Report (WCED 1987). In this perspective, the notion of 'future generations' is vague because it is not easy to define the time horizon of economic, political, moral and strategic choices capable of influencing and conditioning the maintenance of the resources' stock and the conservation of environmental heritage. In fact, the difficulties involved in sustainable development should take into account the time horizon of individual economic agents (e.g. firms, families), which may be restricted to two (or three) generations at most (Pulselli et al. 2008). The issue transitions from individual actions to the organization of agents, specifically various forms of government, ranging from local and regional to national and international levels, which theoretically possess a longer time horizon than individual actors. To address the main issues related to the definition of an interpretative framework of (locally sustainable) development paths, we start from the theoretical and operational definitions of sustainable development – a topic that raises a strong multidisciplinary debate, from environmental, sociological and economic perspectives.

2. Theoretical Foundations of Sustainability

2.1. *Sustainable development: a post-modern concept*

The academic debate on the notion of 'sustainable development' is far from reaching a minimum agreement. One of the major problems encountered concerns the need to introduce qualitative assessments into the definition of 'development' – a concept traditionally analysed using quantitative tools such as Gross Domestic Product (GDP). Foundational contributions by Arrow et al. (1995) and Daly and Cobb (1989) highlighted the inherent tension between economic expansion and environmental limits, emphasizing that conventional models fail to capture the long-term sustainability of socio-ecological systems. These early insights laid the groundwork for rethinking development as a multidimensional concept, involving not only economic indicators but also ecological resilience and human well-being. At this point, two contrasting interpretative approaches to development emerge. On the one hand, the acceptance of a complete cultural 'relativism' denies the possibility of a comparison of the degree of development among countries or regions (e.g. Latouche 1998). On the other hand, searching for (purely quantitative) economic, political and social benchmarks implies that the individual degree of life satisfaction (and quality) can be interpreted as a sort of human development measure, at least indirectly (UNDP 1996).

Between these two extreme positions, several scholars have tried to give theoretical and empirical answers to such questions. Two needs are then faced, i.e. (i) reaching a broader, more flexible and sophisticated concept of sustainable development, and (ii) being able to describe it with empirical analysis' tools. In the face of such challenges, it is no coincidence that the use of per-capita income as a reliable indicator especially capable of representing the degree and dynamics of economic development was predominant in the academic literature (Solow 1993). In a restrictive approach (Arrow et al. 1995), a second variant of the 'constant capital' rule applies, namely the strong sustainability rule. According to this principle, natural capital should remain constant (or increase) within the constraint in which the capital stock remains constant (or increases).

The consequences of introducing the concept of sustainability in the measurement of development have been frequently exemplified (e.g. Palazzi and Lauri 1998). In this perspective, it seems appropriate to use a concept of sustainability related to a working hypothesis assuming all aspects chosen to represent the level of development as fully balanced (Casadio Tarabusi and Palazzi 2004). A balanced development across all components (e.g. economic, social, environmental) provides a valuable foundation for discussions on measuring socio-economic development. This perspective is independent of the inherent characteristics of the variables or indicators used for measurement and allows for the identification of objectives and economic policy mechanisms that integrate sustainability into their goals. The importance of an adequate information – both in terms of coherence and precision in the identification, definition and evaluation of relevant aggregates, and of reliability and timeliness of assessment procedures – becomes one of the inspiring targets for the production of quantitative indicators. This stimulates official statistical producers (usually national or supra-national bodies) to a serious commitment to construction of expanding tools, for instance green accounting, necessary to address a problem so relevant for present and future generations, and joining purely socio-economic aspects with broader (e.g. ecological) considerations.

2.2. *Sustainability, well-being and local development: towards an interpretative framework?*

An academic debate has raised since decades as to whether GDP is an indicator of well-being. Most of the current literature agrees on the need to go beyond it – a sentiment captured in the call for new measures of progress (Costanza et al. 2009) – while less consensus is on how. Increasing the economic product is equivalent to satisfy an increasing number of needs. However, if a vicious circle is established between induced needs and production, the consumption of resources and energy, which are limited, constitutes a barrier to an endless growth. Since GDP origins from the market – where prices are related to the scarcity of goods and services, a rare – even if useless – commodity costs much more than abundant – while useful – one. Price is determined by the marginal utility and not by the total utility, which would be a measure of well-being. Assuming this perspective, does GDP represent something more than (or different from) an economic index? Is it able, albeit roughly, to indicate well-being? If we want to pursue well-being, we cannot simply follow the indications stemming from a purely quantitative and comparative analysis of GDP, e.g. over space and time. Actually, the thesis of a coincidence between GDP and well-being is more popular, but instead theoretically influential, than the one which assumes GDP growth as a necessary precondition for improving well-being. According to this assumption, economic growth would in fact be useful for the environment, the spread of democracy and the decrease in social disparities, among others.

Well-being is, however, multi-systemic. At least, objective and subjective elements exist. Objective well-being is measured through a Quality of Life (QoL) that is objectively measurable, such as based on life expectancy, the degree of violence and crime and the level of corruption. Objective well-being indicators refer to citizens' well-being at the level of countries. The significant developments in measuring subjective well-being (Kahneman and Krueger 2006) have led to proposals for official national well-being accounts (Kahneman et al. 2004), initiatives put into practice by government bodies like the UK's Office for National Statistics (ONS 2025) and through major civil society projects like the Measure of America (MOA 2014). They may seek to combine aspects of a different nature (social, political and cultural ones). The 'subjective' aspects can be found in the affective-emotional sphere and in a dimension of 'emotional stability'. The subjective aspects can be found in the affective-emotional sphere and in a dimension of emotional stability, a field that has been extensively explored by economists and psychologists studying happiness and life satisfaction (Easterlin 1974, 2001; Kahneman 1999; Frey and Stutzer 2002). This multi-systemic view of well-being is particularly critical in contexts of rapid socio-demographic change, where urban expansion and population dynamics fundamentally alter both the physical landscape and the QoL (Véron 2006).

A rethinking of policy goals regulating economic systems means moving from a pure GDP-centred perspective to broader visions centred on well-being (Costanza 2004). The slow deterioration of economic growth prospects results in a latent decline of capitalist societies, which concerns increasingly consistent segments of societies, with a marked imbalance between generations. These aspects become increasingly evident and the incipient economic recession can be interpreted as a crisis of the way of living, producing and consuming. As limits of further growth get more stringent over time, the unsustainability of an infinite expansion of economic activities determines the default of those weaker economies and businesses. The scenario from which to escape is that of an unhappy de-growth, that is of a reduction of well-being combined with a decrease in GDP, which translates into unemployment. Often, even in the presence of a limited increase in per-capita aggregates, there are worse societal conditions that cancel out the slight increase in income.

2.3. *Sustainability, well-being, environmental issues: towards an economic perspective?*

The global depletion of primary natural resources and the negative feedback that this phenomenon causes on biological diversity, climate change and soil degradation, pushes economic theory to propose micro- and macro-level models quantifying the consumption of natural capital and evaluating dynamically the determinants of such phenomena – and their long-term consequences on societies. Nonetheless, many causes, such as structural changes in economic systems, historical and social factors, and the system of public incentives, present controversial effects on land degradation (LD), with temporal and spatial dynamics being still confused and partly explained.

Considering an exquisitely macro-level when analysing economic dynamics, the Environmental Kuznets Curve (e.g. Stern et al. 1996), hereafter EKC, has emerged in the economic literature since the 1990s as a data-driven theory arousing considerable interest, wide debate and vehement criticism in terms of policy implications. This hypothesis links the issue of environmental degradation with the local conditions of economic well-being in which the different states exist, suggesting a non-linear relationship between the two dimensions of territorial complexity (Grossman and Krueger 1995). However, this core relationship is often complicated by other persistent socio-economic factors, such as the 'poverty-environment nexus' that can lock vulnerable rural communities into cycles of unsustainable land management and degradation (Dasgupta and Mäler 1994; Barbier 2010).

Panayotou (1997) provided a summary description of the EKC hypothesis arguing that, at low levels of economic development, the intensity of environmental degradation is constrained by the minimal impact of subsistence activities and by the small quantity of circulating waste. When development increases with (i) the intensification of agriculture, (ii) the take-off of the industrial sector and (iii) a sudden increase in the rate of resource extraction, the rate of resource regeneration drops drastically and the increase in waste – especially toxic ones – becomes significant. However, when the level of economic development reaches extremely high levels, technological innovation, a more effective regulation of production chains at centralized levels and an increased sensitivity of public opinion towards social equality and ecological conservation of natural resources and biodiversity, may determine a gradual decline in environmental degradation.

The EKC relationship is therefore represented by an equation with a quadratic term and therefore takes on an inverted U shape. The maximum point of the curve, usually mentioned as the 'turning point', represents the income level beyond which environmental degradation tends to decrease in the presence of a simultaneous increase in income. The EKC relationship has been widely used in the literature to test the relationship between income and several environmental processes, even when the relationship between the relevant variables is controversial and influenced by exogenous variables of ecological rather than economic nature.

To test the EKC hypothesis, empirical analyses initially used simple linear econometric models through a cross-sectional approach, applied to increasingly larger groups of advanced and emerging countries, in order to highlight the existing dichotomy and the differential impact of income on the level of environmental quality. Subsequently, the refinement of statistical techniques and the improved quality of basic data have allowed the application of panel and time series models, often with specific spatial analysis modules. Some recent studies have also highlighted the usefulness of an inter-country (rather than cross-country) approach, especially where income gaps and ecological levels are different and where the quality of basic data allows approaches at an adequate geographical scale (Vincent 1997). It is clear how, in this analysis, the objective shifts, indirectly, from the study of the implications of the level of economic well-being on the environment and sustainable development, to the effectiveness of the mechanisms for regulating and controlling the negative externalities of the production system in heterogeneous environmental contexts subject to the same regulatory and institutional constraints.

The main criticisms of the EKC hypothesis arose following the often controversial results of the first empirical studies. These controversies cannot be attributed only to the poor quality of statistical data, especially with regard to developing countries. Stern et al. (1996) highlight several critical points of the EKC hypothesis, pointing out the aspect of simultaneity and the influence of international trade, especially with regard to the market of pollution-intensive goods. Arrow et al. (1995) suggest caution in interpreting empirical data in light of the EKC. In particular, the EKC could be valid for pollutants with local effects and costs, not for waste accumulation mechanisms and/or pollutants with spatially dispersed costs and long-term dispersion dynamics. Furthermore, the EKC may ignore the intrinsic dynamics of resource stocks, being mainly interested in pollutant emissions. Finally, although EKC explains the U-shaped relationship that characterizes the various countries analysed through a development path, in which the increase in income elasticity to environmental quality, structural change and policy response, play a dominant role, there are other possible explanations for this trend, among which we should mention the development trajectories that differentially characterize affluent and emerging economies and that do not guarantee uniformity in cross-country analysis.

Despite the caveats that characterize modelling and empirical use of the relationship, the Kuznets curve can be interpreted as a valid logical reference and can suggest economic analysis' tools within a closer integration with simplified interpretative frameworks, where the effectiveness of the response of regulatory (institutional) systems can be evaluated precisely in relation to the income–environment curve. The relationship shows a decreasing trend that highlights how, with high levels of well-being, the negative externalities of the production processes can be mitigated by efficient environmental protection measures, ultimately obtaining an improvement in the level of environmental degradation compared with economically disadvantaged areas. However, in many empirical cases, it should be noted how the relationship tends to lose the typical quadratic trend that characterizes the EKC hypothesis (in the entire range of variation of the function) to assume, instead, a form that tends to be linear for high income values.

3. Frameworks and Indicators for Measuring Sustainability

3.1. *Measuring multiple 'sustainability' dimensions: the DPSIR framework*

One of the most broadly adopted (logical and interpretative) schemes in environmental analysis is the so-called 'Driving forces, Pressures, State, Impacts and Responses' (DPSIR) framework, originally introduced as a 'Pressure-State-Response, PSR' model and further developed by OECD. This scheme has been applied in several empirical studies and aims at highlighting the causal sequence between human pressure, environmental impacts and mitigating responses (OECD 1994). This framework is a reference of European Environment Agency (EEA) for reporting a broad set of environmental issues (EEA 1995). In this scheme, 'driving forces' – which can be conceptualized as both 'fast' and 'slow' variables (Ferrara et al. 2016) – determine environmental issues because of human action and typically include population growth and economic activities such as agriculture, tourism, or industry. The dimensions of 'pressure', 'state' and 'impacts' represent the behaviour of environmental systems (e.g. natural, agricultural) including all known (and latent) interactions among the composing elements. 'Pressures' reflect the direct (and measurable) effect of driving forces (e.g. pollution, urbanization, deforestation). These pressures also include the significant societal and environmental challenges of waste generation and disposal, which have emerged as a critical indicator of unsustainable development paths and spatial injustice (UNEP 2015). The 'state' dimension includes variables that evaluate the (apparent and latent) quality of a given territory in a broad (possibly 'holistic') sense. 'Impacts' are the ultimate effects of the action of specific drivers on a given territory, which make the cause-effect relationships between pressures, state and impacts, immediately explicit and evaluable (e.g. farmland decline, loss of biodiversity, soil degradation, desertification risk).

'Responses' include actions referable to human interventions, such as policies or measures to directly or indirectly contrast environmental degradation (for instance, the common agricultural policy, or the international conventions on desertification, climate change and bio-diversity). In theory, the DPSIR framework aims at highlighting the cause-effect connections specific of a given environmental issue, dynamically evaluating the impact of driving forces as far as sensitivity and capacity for adaptation and response are concerned. DPSIR represents an interpretative reference rather than a methodology in the strict sense for the development of indicators and evaluation procedures. The framework can be operationalized in order to identify specific indicators capable of describing the different components of the model (Salvati et al. 2008).

In this perspective, drivers – consisting of human activities and behaviours deriving from individual and social needs attributable to economic, production and consumption processes – ultimately generate pressures on the environment that alter the state of natural resources. Changes in the state of the environment generate impacts on human health, ecosystems and the economy. Society provides responses that act on the drivers, pressures, state and impacts. The DPSIR framework develops coherently through the adoption of suitable indicators. Referring to the OECD definition (1994), an indicator is

> *a parameter or a value derived from parameters that provides us with information on a phenomenon. The indicator has a meaning that goes beyond the properties that are associated with the value of the parameter. The indicator has a synthetic meaning and is developed for specific purposes.*

The concept of indicators, according to OECD, should not be confused with that of composite (or summary) indexes, which are instead 'a set of weighted parameters or indicators' (OECD 1993; Salvati and Zitti 2009). It should be recalled that there is no 'perfect' indicator to monitor any type of policy, but it is only possible to choose an indicator based on some desired criteria or characteristics. The OECD also highlights some properties of indicators, such as the relevance for the policies adopted (simplicity, rapidity in responding to environmental changes and human activities, offering in turn solid bases for international comparisons and for spatio-temporal comparisons), analytical correctness (reliability, accuracy, precision, reference to international standards) and measurability. The European Environment Agency also underlines the importance of comparability in time and space (Salvati and Zitti 2008). The framework proposed above is therefore described through five classes of indicators:

- 'driver' indicators, providing information on the production sectors and human activities that determine environmental pressures (for instance, industrial production, consumption of goods and services, number of production plants);
- 'pressure' indicators, measuring the cause of a given environmental issue by relating pressures of human systems to the various environmental sectors (for instance, pollutant emissions, consumption of material and energy resources, loss of natural habitats);
- 'state' indicators, estimating the quality of natural systems in both qualitative and quantitative terms (for instance, air quality, water quality, biodiversity);
- 'impact' indicators, assessing the effect of pressures exerted on natural and human systems, such as soil erosion risk under land-use change (Mancino et al. 2016) or damage to human health;

- 'response' indicators, concerning the initiatives undertaken by society and institutional bodies to improve the state of the environment (for instance, laws, plans, funding for scientific research).

In this perspective, non-orthodox economists (e.g. Daly and Cobb 1989; Cobb and Cobb 1994; Neumayer 1999) have proposed measures such as the Index of Sustainable Economic Welfare (ISEW), a concept that has since been refined and debated extensively (Lawn 2003). Technically, ISEW is a summary index that includes some of the components of GDP, incorporating several well-being measures as well, such as income distribution, pollution, infrastructure endowment, car accidents and the degradation of natural capital. Rather than simply adding all expenditures as done to compute GDP, costs and losses due to social and spatial disparities in the distribution of income, depletion of natural resources, environmental degradation and criminal activities, are subtracted from income and two elements of social nature (economic value of leisure time and an approximation of the value of unpaid domestic labour) are added. In other words, ISEW may provide a 'correction' of GDP, while remaining not completely exhaustive. This framework has a rich history of application, with pioneering time-series analyses being constructed for nations like Scotland (Hanley 1999) and some of the earliest work in Italy being conducted in the late 1990s (Guenno and Tiezzi 1998). Since then, the index has been tested in other national and regional contexts, including Belgium (Bleys 2008), other parts of Italy (Gigliarano et al. 2014), and by state-level bodies in the United States of America (MPA 2000). This aligns with a broader international call to move beyond simple economic indicators (Bagstad and Shammin 2009), articulated by the Commission on the Measurement of Economic Performance and Social Progress (Stiglitz et al. 2009). The multidimensionality of well-being requires the construction of a set of additional indicators, at the cost of losing the immediate comparability guaranteed by an individual indicator (Talberth 2008) but ultimately providing a new bottom line for measuring genuine progress (Lawn and Clarke 2006).

4. Empirical Perspectives on Local Sustainable Development

4.1. *Sustainable local development: an interpretative path*

Progress does not happen by chance and is not forever: it should be sustainable and directed towards the search for happiness, avoiding developing societies from the exclusive side of economic growth. With this perspective in mind, the present contribution goes in two directions. In the first part, the issue of sustainability and the problem of analytically quantifying well-being are treated from a theoretical point of view (Quaranta et al. 2020). In the second part, we traced interpretative examples of local development paths under complex background (e.g. territorial) conditions, considering together sustainability, well-being and environmental quality. The discussion will elaborate on the use of GDP as a measure of well-being (Chapter 2) and address its limitations. This will be succeeded by an empirical analysis proposing alternative indicators found in mainstream literature, emphasizing the strengths and weaknesses of both 'objective' and 'subjective' approaches, as well as the empirical research required for their enhancement from both quantitative and qualitative operational perspectives (Chapter 3). An empirical exercise based on official statistics was developed in both chapters covering the same study area and thus allowing for an improved comparability of both points of view.

The transition from theoretical frameworks to practical applications is addressed in the five subsequent empirical chapters, chosen for specific reasons. This book aims to analyse the intricate and practical connections between social systems and physical landscapes that characterize 'sustainable' districts. The chapters are deliberately arranged to construct a thorough, multidimensional argument. This study initiates a discussion on the 'Beyond GDP' debate by first analysing traditional economic well-being and its spatial disparities in Chapter 2, followed by the development of a comprehensive alternative measure for QoL in Chapter 3. Following the establishment of this socio-economic baseline, the subsequent chapters examine key socio-environmental pressure points: the effects of demographic change and urbanization on land (Chapter 4), the relationship between LD and rural poverty (Chapter 5), and the challenge of waste management concerning environmental quality and spatial justice (Chapter 6). These five case studies are selected to offer a comprehensive perspective on the socio-ecological challenges essential for attaining authentic local sustainable development.

In this context, indicators connected to individuals – such as those developed by regional bodies like the Walloon Institute in Belgium (IWEPS 2014) – appear as particularly relevant tools when measuring the level and advances of sustainable development paths. We therefore extended the economic dimension to both ecological and social areas of investigation, considering individual well-being measures. Based on such assumptions, empirical cases were proposed in the subsequent parts of the book in order to represent the full complexity of the intricate relationship between sustainability, well-being and socio-environmental issues in a local development perspective. We specifically addressed exemplificative arguments at the base of any local development path, being representative of the complex nexus between economy, environment and society. They include (i) socio-economic well-being *vis à vis* regional dynamics, (ii) population distribution as a measure of territorial well-being and environmental quality or degradation, (iii) the 'waste' issue as a multifaceted argument stemming from unsustainable development paths with a negative impact on individual and collective well-being levels and (iv) the inexorable expansion of desertification risk in a context of increasing rural poverty. This reflects climate change and land mismanagement in advanced economies, as an indicator of unsustainable – spatially imbalanced and socially unequal – development at various spatial scales, from regional to local (Salvati et al. 2012).

4.2. *Economic growth, spatial disparities and imbalances in well-being levels*

Expansions and recessions were demonstrated to shape socio-demographic inequalities in already divided countries, possibly reducing the industrial performance of advanced economies. However, the specific impact of long waves of growth and prolonged (or recurrent) crises at the local scale was measured only in some countries and for relatively short time intervals. In this perspective, Chapter 2 investigated changes in the distribution of personal incomes across Greek prefectures over a relatively long-time span encompassing 23 years (2000–2022) with the aim at identifying the role of economic downturns and settlement hierarchy in wealth inequalities under both urban and rural conditions. Using statistical tools and spatial techniques, we specifically investigated the geography of per-capita income (Purchasing Power Standard, PPS) considering the distance from the average economic aggregate observed in the European Union.

We developed a purely exploratory framework including descriptive statistics, multivariate techniques, and elementary spatial analysis delineating the role of (i) economic expansion, (ii) the subsequent financial crisis, (iii) industrial recovery and (iv) COVID-19 pandemic in the long-term development path of Greek districts. Based on the adopted indicator, regional clusters displaying homogeneous development paths were identified and discussed explicitly. The empirical outcome of statistical techniques clearly outline how economic expansion paralleled an evident consolidation of the gap between affluent and disadvantaged districts in Greece. The most dynamic performances were observed in metropolitan regions, industrial districts and tourism-specialized areas. The ensuing financial crises demonstrated a gradual decrease in spatial disparities of per-capita income, slightly increasing the gap between metropolitan Athens and the remainder of metropolitan Greece, while the economic performance of rural districts remained largely stable or experienced moderate decline. The empirical results of this study finally suggest how economically disadvantaged, internal and hyper-rural districts are less sensitive to short-term crisis and may respond better to exogenous shocks. The recovery ability of affluent and dynamic districts should be investigated more intensively in the study area and similar contexts, possibly highlighting the geography of socio-economic disparities in highly divided countries and suggesting how expansions and recessions reflect a distinctive distribution of wealth over broader regions.

Moving into the interpretative path from economic to social and environmentally relevant issues, Chapter 3 describes the practical implementation of a composite index assessing QoL at the regional level and points out the main issues that may arise when building up complex frameworks that evaluate regional (sustainable) development paths in advanced economies. The exercise, grounded on a multi-step operational approach grounded on indication theory, was carried out using selected indicators made available from official statistics at the scale of regional administrative domains. To allow generalization to other socio-economic settings in Mediterranean Europe, the investigated context was chosen as reflective of a pre-crisis economic wave of moderate growth and mild social change, the most frequent conditions in the history of the country (and similar European nations) after World War II. Assessing QoL as a basic dimension of sustainable development clarifies the role of individual indicators and composite indexes when describing complex social, economic and demographic phenomena at the base of regional development. By assuming QoL as a relevant dimension of sustainability in divided countries, the empirical results of our study may inform developmental policies in both dynamic and disadvantaged regions of advanced economies.

4.3. *In-between socio-economics and geography: the latent evolution in well-being research*

Rapid urban expansion, originally characterized by dense and relatively compact morphologies, has aroused interest in relation to urban and regional transformations and the decline of traditional landscapes due to an incipient land consumption. In the last century, many urban areas have experienced important changes in both settlement form and economic structure due to urbanization, because of sustained population growth and activities' expansion. In Europe, the compact urbanization typical of the decades immediately following World War II has given way, in recent decades, to looser, branched, inhomogeneous forms of peri-urban settlements, a process that has been documented in detail for major growing cities (e.g. Salvati 2013). They have transformed the ancient patterns of the hamlets and agricultural villages into discontinuous landscapes, with low and (more rarely) medium-density settlements, where service buildings mix with residential ones, fragmenting significantly the agricultural mosaic.

This phenomenon has been attributed mainly to the growth of large cities and the consolidation of networks of medium-sized cities. However, low-density discontinuous settlements, which have expanded rapidly into rural areas, highlight a scale-leap in urbanization processes that increasingly involved pristine and natural land, destroying open spaces, basically a problem that goes beyond urban regions and becomes a 'global' topic, with negative effects on environmental quality and social cohesion (Coluzzi et al. 2022). This trend of urbanization-driven LD in peri-urban areas presents significant socio-economic and environmental challenges throughout Southern Europe (Seifollahi-Aghmiuni et al. 2022). In this perspective, Chapter 4 discusses the implications of urban expansion – intended as a key issue impacting local well-being – that expands into increasingly larger regions, reducing the capacity to recover from external shocks.

More specifically, the study examines Turkey's population dynamics between 2000 and 2021, a period marked by rapid urbanization, demographic dynamics, and economic transformations. Using official data from the Turkish Statistical Institute (TURKSTAT), the analysis explores trends in population growth, internal migration, urbanization and population density across Turkey's provinces. The findings reveal a persistent demographic divide between the urbanized, economically vibrant 'West' and the sparsely populated, economically lagging 'East'. The study highlights two major methodological ruptures that complicate longitudinal analysis: the shift from traditional census methods to the address-based registration system, and the administrative reclassification of rural areas under the 2012 Metropolitan Municipality Law. These changes significantly affect the measurement and interpretation of urbanization and migration patterns.

Despite national population growth, spatial disparities remain entrenched. Western provinces such as Istanbul, Ankara and İzmir continued to attract internal migrants and saw sharp increases in population density, while many eastern provinces experienced sustained out-migration and demographic stagnation. Internal migration flows, although somewhat stabilized over time, continue to reflect and reinforce regional inequalities.

The chapter concludes by emphasizing the policy implications of these demographic patterns, particularly in light of Turkey's aging population and environmental pressures associated with urban sprawl. It calls for a spatially just (and sustainable) approach to regional development that promotes polycentric urbanization, supports rural resilience and prepares for the socio-economic consequences of demographic ageing.

4.4. *Well-being and the environmental challenge*

Chapter 5 discusses soil degradation as an early signal of desertification risk and poverty in economically disadvantaged regions of advanced economies, focusing on case studies exposed to high desertification risk, such as those found in the mountains of Greece (Kosmas et al. 2015, Salvati and Zitti 2005). Research and policy issues in social dynamics and human impacts on soil degradation were finally debated as an original contribution to the study of poverty-environmental nexus in wealthier economies. Environmental infrastructure leads to improvements in social life, strengthening economic development within the scope of sustainable development.

More specifically, the chapter outlines recent findings regarding the intrinsic connections among landscape degradation, desertification risk, social inequalities and rural poverty in developed countries, particularly in Mediterranean Europe. This includes assessing technical strategies for mitigation, such as the '4 per 1000' initiative which focuses on increasing soil organic carbon storage (Francaviglia et al. 2019) and evaluating land vulnerability through process-based approaches (Salvati et al. 2011). The role of key socio-economic factors, including developmental policy, production and market structure, social change and population mobility, in shaping LD was specifically delineated (Salvati et al. 2011). Regional disparities resulting from the intricate interplay of demographic factors (such as urbanization, coastalization and migration) and economic influences (including industrial concentration, crop intensification and tourism pressure) have been characterized by simplified observations from Southern Italy, a region vulnerable to desertification risk (Wilson et al. 2017). With this perspective in mind, the discourse surrounding research and policy matters related to socio-demographic dynamics and human influences on landscape degradation has emerged as a significant contribution to understanding desertification risk in advanced economies (Salvati et al. 2015).

The significant territorial gaps in advanced economies and the different problems on the subject of 'waste management' that emerge at regional scales suggest a more refined analysis of waste management *à* la Kuznets that can make use of quantitatively relevant information, considering homogeneous economic systems. In this context, Chapter 6 makes a specific reference to waste disposal infrastructures, which contribute to an improvement of the environment, as they conduct the flows of materials, from nature, through human activities, back to the environment as waste or emissions. Our analysis addresses this issue by considering advanced economies investigated as independent spatial aggregates, and proposing a targeted elaboration at the level of large geographical areas, in order to capture specializations and territorial peculiarities that would otherwise be latent.

More specifically, the study investigates the persistent North-South divide in Italy's Municipal Solid Waste (MSW) management system, framing it as a critical issue of environmental and spatial injustice that undermines national progress towards sustainable development. Focusing on the formative pre-crisis period of the early 2000s, the analysis investigates the historical and structural roots of this disparity, exploring how uneven infrastructural investment created a path dependency that continues to challenge Italy's transition to a circular economy. The study employs a mixed-methods approach, combining a review of national sustainability policies with a detailed analysis of provincial-level data on waste management practices. A multivariate statistical analysis was adopted to distil complex performance indicators into key underlying factors. The results empirically confirm a stark territorial cleavage, identifying a dominant 'efficiency and modernization' dimension that clearly separates the high-performing Northern provinces from the lagging Southern ones, which remained heavily reliant on landfilling. This infrastructural gap is shown to have locked the regions into divergent socio-economic and environmental trajectories, a problem exacerbated by the influence of organized crime. The study concludes that the North–South divide is not a temporary developmental lag but a structural failure that requires targeted policy interventions (Salvati and Carlucci 2011). A comparative scrutiny of the empirical results allows discussing how a successful and equitable national transition to a circular economy is contingent upon adopting a just transition framework that addresses these deep-seated spatial inequalities through reparative infrastructure investment, empowered local governance and inclusive social dialogue.

5. Discussion

This book proposes a broader academic vision of a mixed (and kaleidoscopic) issue such as the relationship between sustainability, well-being and socio-environmental quality in the perspective of local development. Thanks to vastly articulated and diversified contributions, our contribution mixes economic, social and environmental approaches and perspectives, being decoupled from traditional, disciplinary and academic readings of such relationships, and moving towards a more operational, professional and planning vision of the 'territory' as a key agent of change. In other words, is it preferable to have a high income or more leisure time? Is it better to promote industrialization and employment in a given territory, or preserve its environmental and landscape integrity? Realistically, each of these dimensions is important in defining individual and collective well-being, in which all these aspects should coexist in a truly balanced and sustainable development path at the local scale.

We assume an approach measuring well-being – and possibly, sustainable development, based solely on the economic sphere – as reductive and misleading. Taken as policy-relevant issues, the complexity of the argument at stake requires refined approaches and more elaborate measures than the simple GDP. Only relying on instruments and indicators that involve multiple dimensions of social transformation and economic dynamics, such as the pilot indexes of sustainable economic welfare (Jackson and Marks 1994), it will be possible to conceive economic policies truly sustainable, promoting a net advancement in the QoL of individuals and local communities. Our vision is eminently dynamic and based on both theory and a practical inspection of development paths considering empirical cases grounded on enriched elaborations of different arguments. They were elaborated delineating complex environmental-economic dynamics with important linkages with well-being and local development issues.

Assuming these characteristics as unique in the European literature, at least to our knowledge, this book implements a research and normative perspective that can be easily adapted and generalized to several other contexts, both in advanced economies and in emerging countries, providing a world-wide literature review that can be used as an academic support to sustainable development, applied economics, regional demography, spatial planning, urban-rural sociology, geography and environmental science. The book tries to fill a strong gap in a multidisciplinary perspective dealing with joint arguments of applied economics, sociology and environmental studies at large. Especially the relationship between sustainability, well-being and environmental quality/protection in a local development perspective, in both advanced economies and emerging countries, requires a refined approach and novel interpretative materials and practical lessons, with key examples from local districts. Territories as collective agents of change reflect place-specific, take-home messages which may contribute to scientific theory and academic applications exploring such variegated and multifaceted issues. A clear example of this is the official adoption of policy tools like the Genuine Progress Indicator by state-level governments to guide their planning (Maryland Genuine Progress Indicator 2010).

6. Concluding Remarks

Practical applications at regional and local levels are important, since territories are the collective agent guaranteeing and promoting processes of supra-individual utility referring both to environment quality and social cohesion. These institutions finally promote specific development policies for communities and the sustainable path to be undertaken, with the aim at increasing the effective well-being of individuals. Following such assumptions, and on the basis of recent developments in the scientific literature, we debated extensively on these topics, offering an updated theoretical ground and pragmatic examples. They are all intended at shedding light in an operational evaluation of well-being that integrates the economic dimension with the environmental and social ones, thus assuring a more complete interpretation of complex (local) development paths.

References

Arrow, Kenneth, Bert Bolin, Robert Costanza, Partha Dasgupta, Carl Folke, Crawford S. Holling, Bengt-Owe Jansson, et al. 1995. "Economic Growth, Carrying Capacity and the Environment." *Ecological Economics* 15 (2): 91–95. https://doi.org/10.1016/0921-8009 (95)00059-3.

Bagstad, Kenneth J., and Md Rumi Shammin. 2012. "Can the Genuine Progress Indicator Better Inform Sustainable Regional Progress? – A Case Study for Northeast Ohio." *Ecological Indicators* 18: 330–41. https://doi.org/10.1016/j.ecolind.2011.11.026.

Barbier, Edward B. 2010. "Poverty, Development, and Environment." *Environment and Development Economics* 15 (6): 635–60. https://doi.org/10.1017/S1355770X1000032X.

Bleys, Brent. 2008. "Proposed Changes to the Index of Sustainable Welfare: An Application to Belgium." *Ecological Economics* 64: 741–51. https://doi.org/10.1016/j.ecolecon.2007.10.013.

Casadio Tarabusi, Enrico, and Paolo Palazzi. 2004. "An Index for Sustainable Development." *PSL Quarterly Review* 57 (229): 185–206. https://doi.org/10.13133/2037-3643/9827.

Cobb, Clifford W., John B. Cobb. 1994. *The Green National Product: A Proposed Index of Sustainable Economic Welfare*. Lanham, MD: University Press of America.

Coluzzi, Rosa, Leonardo Bianchini, Gianluca Egidi, Pavel Cudlin, Vito Imbrenda, Luca Salvati, and Maria Lanfredi. 2022. "Density Matters? Settlement Expansion and Land Degradation in Peri-Urban and Rural Districts of Italy." *Environmental Impact Assessment Review* 92: 106703. https://doi.org/10.1016/j.eiar.2021.106703.

Costanza, Robert, Jon Erickson, Karen Fligger, Alan Adams, Christian Adams, Ben Altschuler, Stephanie Balter et al. 2004. "Estimates of the Genuine Progress Indicator (GPI) for Vermont, Chittenden County and Burlington, from 1950 to 2000." *Ecological Economics* 51: 139–55. https://doi.org/10.1016/j.ecolecon.2004.04.009.

Costanza, Robert, Maureen Hart, Stephen Posner, and John Talberth. 2009. *Beyond GDP: The Need for New Measures of Progress*. Pardee Paper no. 4. Boston: Pardee Center for the Study of the Longer-Range Future.

Daly, Herman, and John Cobb, Jr. 1989. *For the Common Good: Redirecting the Economy toward Community, the Environment, and a Sustainable Future*. Boston: Beacon Press.

Dasgupta, Partha and Mäler, Karl-Göran. 1995. "Poverty, institutions, and the environmental resource-base." In *Handbook of Development Economics*, edited by Chenery, Hollis and Srinivasan, T.N. Elsevier, Volume 3, Part A, Pages 2371–2463. https://doi.org/10.1016/S1573-4471(05)80011-7.

Easterlin, Richard A. 1974. "Does Economic Growth Improve the Human Lot? Some Empirical Evidence." In *Nations and Households in Economic Growth: Essays in Honor of Moses Abramovitz*, edited by Paul A. David and Melvin W. Reder, 89–125. New York: Academic Press.

———. 2001. "Income and Happiness: Towards a Unified Theory." *Economic Journal* 111: 465–84. https://doi.org/10.1111/1468-0297.00646.

Ferrara, Agostino, Claire Kelly, Geoff A. Wilson, Angelo Nolè, Giuseppe Mancino, Sofia Bajocco, and Luca Salvati. 2016. "Shaping the Role of 'Fast' and 'Slow' Drivers of Change in Forest–Shrubland Socio-Ecological Systems." *Journal of Environmental Management* 169: 155–66. doi:10.1016/j.jenvman.2015.12.027.

Francaviglia, Rosa, Claudia Di Bene, Roberta Farina, Luca Salvati, and José Luis Vicente-Vicente. 2019. "Assessing '4 per 1000' Soil Organic Carbon Storage Rates under Mediterranean Climate: A Comprehensive Data Analysis." *Mitigation and Adaptation Strategies for Global Change* 24: 795–818. https://doi.org/10.1007/s11027-018-9832-x.

Frey, Bruno S., and Alois Stutzer. 2002. "What Can Economists Learn from Happiness Research?" *Journal of Economic Literature* 40: 402–35. doi:10.1257/002205102320161320.

Gigliarano, Chiara, Francesca Balducci, Maria Teresa Ciommi, and Francesco Chelli. 2014. "Going Regional: An Index of Sustainable Economic Welfare for Italy." *Computers, Environment and Urban Systems* 45: 63–77. https://doi.org/10.1016/j.compenvurbsys.2014.02.007.

Grossman, Gene M., and Alan B. Krueger. 1995. "Economic Growth and the Environment." *Quarterly Journal of Economics* 110 (2): 353–77. https://doi.org/10.2307/2118443.

Guenno, Giorgio, and Silvia Tiezzi. 1998. "*The index of sustainable economic welfare (ISEW) for Italy*." FEEM Working Paper No. 5.98. Fondazione Eni Enrico Mattei (FEEM), Milan.

Hanley, Nick. 1999. "Measuring Sustainability: A Time Series of Alternative Indicators for Scotland." *Ecological Economics* 28: 55–73. https://doi.org/10.1016/S0921-8009(98)00027-5.

IWEPS (Walloon Institute for Evaluation, Prospective and Statistics). 2014. *Indicators Complementary to GDP: Index of Conditions of Well-Being in Wallonia (ICWB), April 2014.*

Jackson, Tim, Nic Marks, and Stockholm Environment Institute. 1994. *Measuring sustainable economic welfare: a pilot index: 1950–1990.* Stockholm Environment Institute.

Kahneman, Daniel. 1999. "Objective Happiness." In *Well-Being: Foundations of Hedonic Psychology*, edited by Daniel Kahneman, Ed Diener, and Norbert Schwarz, 3–25. Russell Sage Foundation.

Kahneman, Daniel, and Alan B. Krueger. 2006. "Developments in the Measurement of Subjective Well-Being." *Journal of Economic Perspectives* 20 (1): 3–24. doi: 10.1257/089533006776526030.

Kahneman, Daniel, Alan B. Krueger, David Schkade, Norbert Schwarz, and Arthur Stone. 2004. "Toward National Well-Being Accounts." *American Economic Review, Papers and Proceedings* 94: 429–34. doi: 10.1257/0002828041301713.

Kosmas, Costas, Vassilis Detsis, Maria Karamesouti, Konstantina Kounalaki, Panayiotis Vassiliou, and Luca Salvati. 2015. "Exploring Long-Term Impact of Grazing Management on Land Degradation in the Socio-Ecological System of Asteroussia Mountains, Greece." *Land* 4 (3): 541–59. doi:10.3390/land4030541.

Latouche, Serge. 1998. *Les Dangers du Marché Planétaire.* Paris: Presses de Sciences Po.

Lawn, Philip. 2003. "A Theoretical Foundation to Support the Index of Sustainable Economic Welfare (ISEW), Genuine Progress Indicator (GPI) and Other Related Indexes." *Ecological Economics* 44: 105–18.

Lawn, Philip and Matthew Clarke. 2006. "Comparing Victoria's Genuine Progress with That of the Rest of Australia." *Journal of Economic and Social Policy* 10 (2): 115–38.

Lewis, Kristen, and Sarah Burd-Sharps. 2014. *American human development report: The measure of America 2013–2014.* Measure of America of the Social Science Research Council.

Mancino, Giovanni, Alessandra Nolè, Luca Salvati, and Antonio Ferrara. 2016. "In-Between Forest Expansion and Cropland Decline: A Revised USLE Model for Soil Erosion Risk under Land-Use Change in a Mediterranean Region." *Ecological Indicators* 71: 544–50. https://doi.org/10.1016/j.ecolind.2016.07.040.

Maryland Genuine Progress Indicator. 2010. *An Index for Sustainable Prosperity*. Annapolis, MD: Maryland Department of Natural Resources. https://dnr.maryland.gov/mdgpi/Pages/default.aspx.

Minnesota Planning Agency (MPA). 2000. *Smart Signals: An Assessment of Progress Indicators*. Retrieved June 12, 2025, from http://www.ecologydemocracynetwork.org/downloads/EQB_Smart_Signals_1990_measure.pdf.

Neumayer, Eric. 1999. "The ISEW: Not an Index of Sustainable Economic Welfare." *Social Indicators Research* 48, no. 1: 77–101. http://www.jstor.org/stable/27522403.

OECD (Organisation for Economic Co-operation and Development). 1993. *OECD Core Set of Indicators for Environmental Performance Reviews*. Environment Monographs No. 83. Paris: OECD.

OECD (Organisation for Economic Co-operation and Development). 1994. *Environmental Indicators: OECD Core Set*. Paris: OECD.

ONS (Office for National Statistics). 2025. *UK Measures of National Well-Being User Guide*. Last revised February 11, 2025. London: ONS. https://www.ons.gov.uk/.../ukmeasuresofnationalwellbeinguserguide.

Palazzi, Paolo, and Agnese Lauri. 1998. The Human Development Index: suggested corrections. *PSL Quarterly Review*, 51 (205). https://doi.org/10.13133/2037-3643/10596.

Panayotou, Theodore. 1997. "Demystifying the Environmental Kuznets Curve: Turning a Black Box into a Policy Tool." *Environment and Development Economics* 2 (4): 465–84. http://www.jstor.org/stable/44379189.

Pulselli, Federico Maria, Simone Bastianoni, Nadia Marchettini, and Enzo Tiezzi. 2008. *The Road to Sustainability: GDP and Future Generations*. Southampton, UK: WIT Press.

Quaranta, Giovanni, Rosanna Salvia, Luca Salvati, Valentina Di Paola, Raffaele Coluzzi, Valerio Imbrenda, and Tiziana Simoniello. 2020. "Long-Term Impacts of Grazing Management on Land Degradation in a Rural Community of Southern Italy: Depopulation Matters." *Land Degradation & Development* 31 (16): 2379–94. https://doi.org/10.1002/ldr.3583.

Salvati, Luca. 2013. "Monitoring High-Quality Soil Consumption Driven by Urban Pressure in a Growing City (Rome, Italy)." *Cities* 31: 349–56.

Salvati, Luca, and Margherita Carlucci. 2011. "The Economic and Environmental Performances of Rural Districts in Italy: Are Competitiveness and Sustainability Compatible Targets?" *Ecological Economics* 70 (12): 2446–53. doi:10.1016/j.ecolecon.2011.07.030.

Salvati, Luca, and Marco Zitti. 2005. "Land Degradation in the Mediterranean Basin: Linking Bio-Physical and Economic Factors into an Ecological Perspective." *Biota* 6: 67–77.

———. 2008. "Regional Convergence of Environmental Variables: Empirical Evidences from Land Degradation." *Ecological Economics* 68 (1–2): 162–68. https://doi.org/10.1016/j.ecolecon.2008.02.018.

———. 2009. "Substitutability and Weighting of Ecological and Economic Indicators: Exploring the Importance of Various Components of a Synthetic Index." *Ecological Economics* 68 (4): 1093–99. https://doi.org/10.1016/j.ecolecon.2008.07.017.

Salvati, Luca, Sofia Bajocco, Tomaso Ceccarelli, Marco Zitti, and Luigi Perini. 2011. "Towards a Process-Based Evaluation of Land Vulnerability to Soil Degradation in Italy." *Ecological Indicators* 11 (5): 1216–27. https://doi.org/10.1016/j.ecolind.2010.12.024.

Salvati, Luca, Anastasios Mavrakis, Andrea Colantoni, Giuseppe Mancino, and Agostono Ferrara. 2015. "Complex Adaptive Systems, Soil Degradation and Land Sensitivity to Desertification: A Multivariate Assessment of Italian Agro-Forest Landscape." *Science of the Total Environment* 521: 235–45. https://doi.org/10.1016/j.scitotenv.2015.03.094.

Salvati, Luca, Vittorio Gargiulo Morelli, Kostas Rontos, and Alberto Sabbi. 2013. "Latent Exurban Development: City Expansion along the Rural-to-Urban Gradient in Growing and Declining Regions of Southern Europe." *Urban Geography* 34 (3): 376–94. https://doi.org/10.1080/02723638.2013.778675.

Salvati, Luca, Luigi Perini, Alberto Sabbi, and Sofia Bajocco. 2012. "Climate Aridity and Land Use Changes: A Regional-Scale Analysis." *Geographical Research* 50 (2): 193–203. https://doi.org/10.1111/j.1745-5871.2011.00723.x.

Salvati, Luca, Flavia Ranalli, Margherita Carlucci, Achille Ippolito, Antonio Ferrara, and Piermaria Corona. 2017. "Forest and the City: A Multivariate Analysis of Peri-Urban Forest Land Cover Patterns in 283 European Metropolitan Areas." *Ecological Indicators* 73: 369–77. https://doi.org/10.1016/j.ecolind.2016.09.025.

Salvati, Luca, Ilaria Tombolini, Luigi Perini, and Agostino Ferrara. 2013. "Landscape Changes and Environmental Quality: The Evolution of Land Vulnerability and Potential Resilience to Degradation in Italy." *Regional Environmental Change* 13: 1223–33. https://doi.org/10.1007/s10113-013-0437-3.

Salvati, Luca, Giuseppe Venanzoni, Pere Serra, and Margherita Carlucci. 2016. "Scattered or Polycentric? Untangling Urban Growth in Three Southern European Metropolitan Regions through Exploratory Spatial Data Analysis." *The Annals of Regional Science* 57: 1–29. https://doi.org/10.1007/s00168-016-0758-5.

Salvati, Luca, Marco Zitti, and Tomaso Ceccarelli. 2008. "Integrating Economic and Environmental Indicators in the Assessment of Desertification Risk: A Case Study." *Applied Ecology and Environmental Research* 6 (1): 129–38.

Seifollahi-Aghmiuni, Shirin, Zahra Kalantari, Gianluca Egidi, Luisa Gaburova, and Luca Salvati. 2022. "Urbanisation-Driven Land Degradation and Socioeconomic Challenges in Peri-Urban Areas: Insights from Southern Europe." *Ambio* 51 (6): 1446–58. doi:10.1007/s13280-022-01701-7.

Solow, Robert. 1993. "An Almost Practical Step toward Sustainability." *Resources Policy* 19 (3): 162–72. https://doi.org/10.1016/0301-4207(93)90001-4.

Stanners, David, and Philippe Bourdeau, eds. 1995. *Europe's Environment: The Dobříš Assessment.* Luxembourg: Office for Official Publications of the European Communities.

Stern, David I., Michael S. Common, and Edward B. Barbier. 1996. "Economic Growth and Environmental Degradation: The Environmental Kuznets Curve and Sustainable Development." *World Development* 24 (7): 1151–60. https://doi.org/10.1016/0305-750X(96)00032-0.

Stiglitz, Joseph E., Amartya Sen, and Jean-Paul Fitoussi. 2009. *Report by the Commission on the Measurement of Economic Performance and Social Progress.* Paris: Commission on the Measurement of Economic Performance and Social Progress.

Talberth, John. 2008. "A New Bottom Line for Progress." *In State of the World 2008: Ideas and Opportunities for Sustainable Economies (25th ed.).* Routledge. https://doi.org/10.4324/9781849772600.

UNDP (United Nations Development Programme). 1996. *Human Development Report 1996: Economic Growth and Human Development.* New York: Oxford University Press.

UNEP (United Nations Environment Programme). 2015. *Global Waste Management Outlook.* Report. ISBN 978-92-807-3479-9. https://eprints.whiterose.ac.uk/id/eprint/99773/.

Véron, René. 2006. Remaking Urban Environments: The Political Ecology of Air Pollution in Delhi. *Environment and Planning A: Economy and Space*, 38 (11), 2093–2109. https://doi.org/10.1068/a37449.

Vincent, Jeffrey R. 1997. "Resource Depletion and Economic Sustainability in Malaysia." *Environment and Development Economics* 2 (1): 19–37. https://doi.org/10.1017/S1355770X97000107.

Wilson, Geoff A., Clare L. Kelly, Helen Briassoulis, Agostino Ferrara, Giovanni Quaranta, Rosanna Salvia, Vassilis Detsis, et al. 2017. "Social Memory and the Resilience of Communities Affected by Land Degradation." *Land Degradation & Development* 28 (2): 383–400. https://doi.org/10.1002/ldr.2669.

WCED (World Commission on Environment and Development). 1987. *Brundtland Report: Our Common Future.* New York: United Nations. https://www.brundtland.co.za/other-publication/brundtland-report-1987-our-common-future/.

Chapter 2

ECONOMIC DOWNTURNS, INCOME INEQUALITIES AND SETTLEMENT HIERARCHIES: UNDERSTANDING THE GREEK REGIONAL DEVELOPMENT PATH, 2000–2022

Bünyamin Özkür, Marco Maialetti, Donato Scarpitta, Adele Sateriano, Pierpaolo Lombardi, Enrico Maria Mosconi, Luca Salvati and Ioannis Konaxis

1. Introduction

Convergence (or divergence) between macro-regions stimulated an ever-growing academic debate on spatial disparities, specifically focusing on the causes underlying social inequalities and demographic divides (Storper 2011; Glaeser et al. 2012; Martin and Sunley 2014). Reflecting a positive challenge in urban studies and regional science, applied economics has given some possible explanations for unequal processes of regional growth – both looking at long-term dynamics, or basically focusing on medium (or even shorter) times (Barro and Sala-i-Martin 2003). In this perspective, convergence mechanisms resulted in partial reductions of social inequalities in both affluent and disadvantaged economic systems (Arbia and Paelinck 2003; Ezcurra et al. 2007). However, empirical research – while ascertain constant returns to scale and declining capital productivity – has documented similar and converging levels of per-capita income in close regions only in some cases (Henning et al. 2013). Comparative exercises dealing with unemployment rates, investment levels, innovation dynamics, among others, have instead showed less converging findings (Dixon et al. 2001; Liao and Wei 2012; Fratesi and Percoco 2013).

Convergence in the level of income and wealth across regions and districts is an academic issue of increasing interest since the unification of member states into European Union (Liberati 2015). In this view, reducing regional inequalities has become a foundational goal of European cohesion policy (King et al. 2000; Van Well 2012; ESPON 2014). Despite significant policy attention, long-standing territorial imbalances persist, particularly in Southern Europe (Leontidou and Marmaras 2001). Notable socio-economic gaps continue to characterize many inland and marginal areas in countries such as Portugal, Spain, Italy, Greece and Turkey (Sicular et al. 2007). These disparities remain entrenched despite large-scale public investments and institutional reforms. Scholars have pointed to various underlying causes, including the configuration of housing markets, state-market interactions, urban hierarchies, welfare models and the evolving role of national institutions (Terrasi, 1999; Cracolici et al. 2007; Lois, 2009; Kyriakou and Roca-Sagales 2014). Yet, sluggish changes in productive structures, human capital limitations, and depopulation trends continue to reinforce the divide between dynamic metropolitan zones – often coastal – and stagnant inland regions (Prodromidis 2014).

While long-term convergence remains a widely studied topic, short-term variations in regional performance have attracted far less scholarly attention (Duran 2014). Many empirical studies of territorial inequality overlook the role of sudden shocks – such as financial crises, recessions, or policy disruptions – on localized development paths (Frenken et al., 2007; Chapple and Lester 2010; Pendall et al. 2010; Martin, 2012). These shocks can trigger rapid economic divergence, depending on the structural features of a region. Fingleton et al. (2012) emphasized that regions differ significantly in their ability to withstand and recover from such downturns. For instance, Frenken et al. (2007) argued that, while specialization may support medium-term employment growth, it increases exposure to volatility. In the United States, regional differences in the timing of cyclical fluctuations, rather than the magnitude, explain much of the short-term divergence in income levels (Duran 2014). Rodríguez-Pose and Fratesi (2007) found that regional disparities evolved pro-cyclically in countries like Italy, Portugal, and Spain – but not in Greece or France – due to structural characteristics such as the existence of sheltered economies.

The global financial crisis of 2008–2009 profoundly altered regional economic trajectories across Europe (Lee 2012). The differential impact of this shock prompted a surge in studies examining regional resilience – the capacity of local economies to absorb and rebound from downturns (Kemeny and Storper 2014). Scholars observed that regions with greater sectoral diversity tended to experience more stable outcomes, even if their peak growth was slower (Essletzbichler 2007). Diversity in the industrial structure enhances adaptive capacity and buffers against sector-specific collapse. Holm and Østergaard (2013) note that regions with large firms often better endure economic shocks, while those with high rates of start-ups may be more flexible in post-crisis recovery.

At the same time, what has been termed a 'resilience gap' can emerge, where regions differ not only in economic structure, but also in institutional capacity, labour market adaptability and demographic characteristics. This gap can reinforce regional disparities in unemployment and income (Christopoulos 2004; Patacchini and Rice 2007; Taulbut and Robinson 2014). In the United Kingdom, Lee (2012) found that cities with higher proportions of skilled employment in finance, construction and manufacturing, recorded smaller increases in unemployment during the crisis. Overall, however, the evidence across European countries remains inconclusive, with studies revealing divergent patterns due to highly localized factors (Davies 2011).

Within this economic landscape, Greece presents a particularly instructive case (Tsakloglou 1990). The country experienced one of the deepest and most prolonged recessions in the Eurozone. Despite the temporary economic boost from the 2004 Olympic Games, Greece remained structurally reliant on construction, commerce, and light manufacturing – sectors that proved highly vulnerable to external shocks. By 2014, national unemployment exceeded 26%, with youth unemployment surpassing 50%. These conditions significantly hindered efforts to diversify the economy or modernize regional development trajectories.

As such, Greece offers a valuable opportunity to investigate how regional systems respond to short-run economic fluctuations (e.g. Cho et al. 2014). With this perspective in mind, the present chapter analyses the spatial and temporal evolution of per capita income in Greek prefectures from 2000 to 2022. Using a combination of exploratory data analysis, multivariate statistics and spatial econometric techniques (Patacchini 2008), we examine the links between district-level economic performance and broader patterns of inequality. Our approach emphasizes the importance of local territorial characteristics – such as institutional capacity, sectoral diversity and demographic trends – in shaping regional resilience.

In doing so, we contribute to the literature on regional resilience and inequality by highlighting the dual role of short-run shocks in both exacerbating and, at times, reconfiguring spatial disparities. Changes in per capita income are interpreted as indicators of either resistance to downturns (i.e. limited income reduction) or recovery following shocks (i.e. significant income gains). Moreover, we connect these patterns to Greece's urban-rural structure, offering insights into the geography of socio-economic resilience. This chapter thus informs place-based policy design and fosters more equitable regional development.

2. Methodology

2.1. *Study area*

The analysis focuses on the entirety of Greece, which spans a land area of 131,982 km^2. The elementary observation unit is the ensemble of 51 prefectures ('nomoi' in Greek), which correspond to the NUTS-3 level within the European Union's statistical territorial hierarchy. Positioned between municipalities and broader regional administrations, these prefectures encapsulate a wide array of socio-economic conditions and provide a robust spatial framework for evaluating developmental dynamics. Their granularity also enables the exploration of spatial gradients such as elevation, proximity to urban centres and coastal–inland contrasts.

The Greek urban system is highly centralized, with over 30% of the national population concentrated in Attica, the administrative region including metropolitan Athens – home to more than three million inhabitants. Beyond Athens, several medium-sized urban centres, such as Thessaloniki and Heraklion, play key economic roles, alongside provincial capitals like Patras, Larissa, Volos, Kalamata and Chania. However, the distribution of population and economic activity is uneven. While coastal regions and islands in the Aegean and Ionian Seas benefit from tourism-driven growth, interior rural areas often suffer from demographic decline and structural economic limitations. The spatial heterogeneity across these prefectures provides a valuable basis for understanding how regional development is shaped by geography, administrative capacity and sectoral specialization.

2.2. *Data sources and variables*

This study relies on a harmonized time series of annual per capita gross income data (in Euros), published by the Hellenic Statistical Authority (ELSTAT) for Greek prefectures (NUTS-3 level) since the year 2000. The dataset follows the standards of the European System of Accounts (ESA 2010), with 2010 serving as the benchmark year, in accordance with European Commission Regulation EC 549/2013. To enhance cross-regional comparability, income values have been adjusted to reflect regional price differences using PPS, as provided by Eurostat. This adjustment allows for an accurate depiction of income levels in real terms across both Greek regions and the broader European context over time.

In addition, annual demographic data at the NUTS-3 level have been sourced from ELSTAT, based on both population census data and the national civil registry. These figures help contextualize income patterns within broader demographic trends such as population density, ageing and urbanization.

In this work, after adapting regional GDP figures to PPS (EU27 from 2020) per inhabitant, we calculated it as a percentage of the EU27 (from 2020) average. This metric may highlight – more clearly than other simplified approaches – the socio-economic distance between Greek regions and the rest of the old continent, and their evolution over a so long time interval (2000–2022).

2.3. *Descriptive metrics: methodological framework*

To identify groups of regions exhibiting similar developmental trajectories, we conducted a detailed statistical analysis on a standardized data matrix of per capita income values (adjusted using PPS). These values represent the deviation of each Greek prefecture's income from the annual average across European Union (EU) member states, covering the time interval from 2000 to 2022. The analysis provides insights into spatial and temporal trends in economic inequality and regional disparities in Greece over a prolonged timeframe.

The descriptive framework incorporates traditional statistical measures – such as the arithmetic and geometric means, median, minimum and maximum values – as well as dispersion indicators, including the coefficient of variation, skewness and kurtosis. These metrics offer a comprehensive understanding of central tendency, distribution shape, and volatility within the per capita income data. To deepen the analysis, we also included a set of more innovative indicators, first proposed by Salvati et al. (2014), which are particularly suited to tracking changes in the spatial configuration of socio-economic variables. These include:

- median-to-mean ratio, which serves as a basic indicator of skewness or asymmetry;
- normalized range, calculated using the minimum, maximum and arithmetic mean values;
- percentile range, based on the interquartile range (i.e. 25th–75th percentile) and median.

To further explore the heterogeneity in income distributions between affluent and disadvantaged regions, we computed a suite of diversity and inequality metrics inspired by information theory and ecological science. These indicators capture both the concentration and evenness of income levels across prefectures, as follows:

- Simpson's Index (S), estimating dominance in the income distribution; a value near 0 indicates uniform income levels, while values approaching 1 suggest strong concentration in specific areas;

- Shannon Index (*H*), reflecting the degree of diversification in income levels, with higher values denoting more equitable distributions;
- Brillouin's Index (*B*), complementing Shannon's metric by accounting for observed combinations of income ranks;
- Margalef's and Fisher-Alpha indices, offering alternative views of diversity within the distribution, particularly regarding income variability;
- Berger–Parker index, assessing the degree to which the highest income value dominates the overall distribution;
- Menhinick's Richness Index (*M*), quantifying statistical heterogeneity based on the ratio of income categories to sample size;
- Pielou's Evenness Index (calculated as the Shannon Index divided by the logarithm of the number of prefectures), evaluating the uniformity of income distribution across space.

Together, these descriptive statistics provide a robust framework for interpreting the temporal evolution and spatial patterns of economic disparities in Greece. They also allow for a multidimensional assessment of how income inequality varies across different regions and time periods.

2.4. *Multivariate analysis*

To explore the underlying structure of regional income disparities across Greece, a Principal Component Analysis (PCA) was conducted on the absolute values of the twelve descriptive indicators previously outlined. The PCA was applied to the entire temporal dataset – spanning all observation years (2000–2022) and spatial domains (NUTS-3 prefectures). This multivariate approach serves two primary purposes: to reduce redundancy among correlated indicators and to identify patterns of co-variation that reveal the latent dimensions shaping the spatial dynamics of per capita income across regions (Anselin 2001).

The PCA was computed using the correlation matrix to ensure comparability across indicators with different units and variances. In selecting the number of meaningful components, we applied the common criterion of retaining those with eigenvalues greater than 1 (Coppi et al. 1989). These components represent distinct, interpretable axes of variation and provide a compact summary of the multidimensional structure of income inequality in Greece.

To validate the suitability of the dataset for PCA, we used the Kaiser–Meyer–Olkin (KMO) measure of sampling adequacy, which assesses whether the observed variables exhibit sufficient shared variance. Additionally, Bartlett's test of sphericity was employed to determine whether the correlation matrix significantly differs from an identity matrix – an essential condition for reliable

PCA results (Salvati et al. 2013). High KMO values and statistically significant Bartlett's test results indicated the robustness of our PCA model.

Following the extraction of principal components, we applied a hierarchical cluster analysis to classify prefectures into homogeneous groups. This clustering was based on Ward's agglomerative algorithm, which minimizes the total within-cluster variance and utilized Euclidean distances as the dissimilarity metric. The combination of PCA and hierarchical clustering enabled a multidimensional characterization of regional economic profiles and provided a more nuanced understanding of spatio-temporal disparities in income levels.

3. Results

3.1. *Descriptive statistics*

The spatial distribution of per-capita income in Greek was illustrated in Table 1. Per-capita income distributed along the geographical gradient from Attica (including the capital city of Athens) to more peripheral districts in Northern Greece, with less intense disparities in Northern Greece, when considering the possible divide between Thessaloniki, the second Greek city, and the surrounding regions. The number of prefectures with per-capita income higher than the European average maintained relatively low in the wave of economic expansion (2000–2007), decreasing drastically afterwards, and consolidating the spatial gap between marginal rural areas and coastal districts. While concentrating the highest income in recent times, metropolitan areas (both Athens and Thessaloniki) and more accessible internal lowlands specialized in manufacturing and high-input agriculture (e.g. Viotia) experienced a strong decline in the level of personal wealth. More specifically, the districts around Attica or with tourism specialization (e.g. Dodecannese islands, Ionian islands, Evia, Kyklades) were areas showing per-capita income persistently higher (or aligned with) the European average only up to 2007. Crisis has heavily impacted these districts, likely more intensively than the other Greek areas, despite their lower (initial) level of wealth. Conversely, a persistently low per-capita income was observed in less accessible prefectures of Northern and Western Greece, with a relatively stable ranking compared with European average.

3.2. *Spatial disparities in per-capita income: ecological metrics*

Tables 2, 3, and 4 present a comprehensive set of descriptive and diversity metrics calculated for income values (in PPS-adjusted Euros) across Greek prefectures over three distinct phases: the pre-crisis economic expansion

Table 1. Descriptive statistics of standardized per-capita income (Euros) across Greek prefectures during economic expansion, by year (PPS, EU27 from 2020, per inhabitant in percentage of the EU27 (from 2020) average).

Year	Greece	Attica	Northern Aegean	Southern Aegean	Crete	Eastern Macedonia, Trace	Central Macedonia	Western Macedonia	Epyrus	Thessalia	Ionia Nisia	Western Greece	Central Greece	Peloponnese
2000	85	108	61	99	77	66	70	78	65	62	85	65	92	72
2001	88	111	64	99	81	67	72	80	67	65	89	67	93	75
2002	91	117	63	96	83	68	74	79	69	71	86	69	93	75
2003	94	121	71	102	85	68	75	84	70	76	93	72	96	77
2004	96	124	71	105	88	69	78	85	68	76	94	74	93	76
2005	93	121	71	104	84	67	74	83	66	71	94	72	92	75
2006	96	128	74	106	86	65	76	85	67	73	94	75	90	78
2007	93	123	73	102	81	65	74	78	63	70	91	72	85	75
2008	93	124	74	105	82	66	75	76	62	70	93	71	85	76
2009	94	126	75	102	82	67	75	79	62	70	89	69	83	76
2010	84	114	66	91	73	63	66	76	58	61	80	64	74	68
2011	74	101	58	80	62	53	59	72	52	53	67	55	66	61
2012	70	94	55	75	57	50	55	80	47	52	63	53	63	59
2013	71	96	56	81	60	49	55	80	49	53	65	53	63	60
2014	71	96	56	81	61	48	54	81	48	53	67	53	61	59

2015	69	94	53	77	60	47	54	73	47	52	64	52	60	59
2016	67	92	50	72	57	46	53	66	46	51	62	50	59	58
2017	67	91	48	71	57	45	52	65	45	50	61	49	59	58
2018	66	91	45	71	56	45	53	63	45	50	61	48	59	57
2019	66	91	43	70	56	43	52	58	44	50	61	47	58	56
2020	62	86	39	57	49	42	49	53	41	48	50	45	58	53
2021	64	89	40	61	52	43	50	54	42	51	54	46	59	54
2022	67	92	41	65	57	44	54	56	43	51	58	49	63	59
2023	69	96	42	70	59	45	55	55	45	52	61	50	63	59

Source: Authors' elaboration on ELSTAT and Eurostat data.

Table 2. Descriptive statistics of per-capita income (Euros) across Greek prefectures during economic expansion, by year.

Metric	2000	2001	2002	2003	2004	2005	2006	2007
Minimum	51	53	53	54	56	53	51	51
Maximum	136	140	149	168	169	156	183	167
Arith. mean	77.1	79.3	80.7	84.0	85.2	82.6	84.2	81.2
Median	71.5	72.0	76.5	80.5	79.0	77.5	79.0	75.5
25th percentile	62.0	64.3	64.3	65.5	68.0	66.3	65.8	65.0
75th percentile	88.0	89.5	87.8	92.8	93.8	89.8	90.5	86.5
Skewness	1.3	1.3	1.5	1.6	1.5	1.4	1.8	1.7
Kurtosis	1.3	1.6	2.4	2.7	2.6	1.8	3.7	3.2
Geom. mean	74.8	76.9	78.3	81.2	82.4	79.8	80.9	78.2
Coeff. Variation	26.6	26.5	27.2	28.7	28.7	28.2	31.3	30.3
Median-to-mean	0.927	0.909	0.948	0.958	0.927	0.938	0.938	0.930
Normalized range	1.10	1.10	1.19	1.36	1.33	1.25	1.57	1.43
Percentile range	0.36	0.35	0.31	0.34	0.33	0.30	0.31	0.28
Simpson	0.979	0.979	0.979	0.979	0.979	0.979	0.979	0.979
Shannon	3.919	3.920	3.918	3.915	3.915	3.915	3.908	3.911
Evenness	0.968	0.969	0.968	0.964	0.964	0.965	0.958	0.961
Brillouin	3.880	3.882	3.881	3.879	3.879	3.879	3.873	3.874
Menhinick	0.821	0.810	0.803	0.787	0.781	0.794	0.786	0.800
Margalef	6.147	6.127	6.114	6.084	6.074	6.097	6.083	6.109
Fisher-alpha	8.432	8.388	8.359	8.295	8.273	8.322	8.292	8.349
Berger–Parker	0.034	0.034	0.036	0.038	0.038	0.036	0.042	0.040

Source: Authors' elaboration on ELSTAT and Eurostat data.

Table 3. Descriptive statistics of per-capita income (Euros) across Greek prefectures during recession, by year.

Metric	2008	2009	2010	2011	2012	2013	2014	2015
Minimum	50	50	47	42	42	43	41	41
Maximum	163	167	151	135	128	132	135	133
Arith. mean	81.5	81.3	73.3	64.5	61.2	62.2	62.0	60.1
Median	77.0	76.5	67.0	59.0	56.0	56.0	56.5	54.5
25th percentile	65.0	64.3	58.0	52.0	49.0	49.0	48.3	48.0
75th percentile	84.8	86.0	78.8	69.0	66.5	69.0	69.5	67.8
Skewness	1.7	1.8	1.8	1.9	1.7	1.7	1.7	1.9

Table 3. (Continued)

Kurtosis	3.2	3.6	3.5	3.6	3.0	3.1	3.5	4.6
Geom. mean	78.6	78.4	70.8	62.2	59.0	59.9	59.6	58.0
Coeff. Variation	30.1	30.3	29.9	30.3	30.0	30.5	31.0	30.3
Median-to-mean	0.945	0.941	0.914	0.915	0.915	0.900	0.912	0.906
Normalized range	1.39	1.44	1.42	1.44	1.40	1.43	1.52	1.53
Percentile range	0.26	0.28	0.31	0.29	0.31	0.36	0.38	0.36
Dominance	0.0209	0.0210	0.0209	0.0210	0.0209	0.0210	0.0210	0.0210
Simpson	0.979	0.979	0.979	0.979	0.979	0.979	0.979	0.979
Shannon	3.912	3.911	3.912	3.912	3.912	3.911	3.909	3.912
Evenness	0.961	0.961	0.962	0.961	0.961	0.960	0.959	0.961
Brillouin	3.875	3.874	3.872	3.867	3.865	3.865	3.863	3.864
Menhinick	0.799	0.800	0.842	0.898	0.922	0.914	0.916	0.930
Margalef	6.106	6.108	6.185	6.283	6.323	6.310	6.314	6.337
Fisher-alpha	8.343	8.348	8.515	8.736	8.828	8.799	8.806	8.860
Berger–Parker	0.038	0.040	0.040	0.040	0.040	0.041	0.042	0.043

Source: Authors' elaboration on ELSTAT and Eurostat data.

Table 4. Descriptive statistics of per-capita income (Euros) across Greek prefectures during post-crisis (pre-pandemic) economic recovery and COVID-19 pandemic, by year.

Metric	2016	2017	2018	2019	2020	2021	2022
Minimum	39	38	38	37	35	37	38
Maximum	132	133	132	130	124	129	132
Arith. mean	58.2	57.4	56.8	56.1	52.2	53.9	57.0
Median	52.5	52.0	51.0	50.5	47.0	49.5	52.0
25th percentile	47.0	46.0	45.0	43.3	41.0	43.0	44.0
75th percentile	64.0	63.5	62.8	61.5	54.8	58.0	61.0
Skewness	2.1	2.1	2.1	2.0	2.3	2.3	2.1
Kurtosis	5.4	5.6	5.5	5.1	6.2	6.4	5.1
Geom. mean	56.1	55.2	54.7	53.9	50.1	51.7	54.7
Coeff. Variation	30.7	31.8	31.8	32.5	33.4	33.4	32.9
Median-to-mean	0.902	0.906	0.897	0.900	0.900	0.919	0.912

(*continued*)

Table 4. (Continued)

Metric	2016	2017	2018	2019	2020	2021	2022
Normalized range	1.60	1.66	1.65	1.66	1.70	1.71	1.65
Percentile range	0.32	0.34	0.35	0.36	0.29	0.30	0.33
Dominance	0.0210	0.0211	0.0211	0.0212	0.0213	0.0213	0.0213
Simpson	0.979	0.979	0.979	0.979	0.979	0.979	0.979
Shannon	3.911	3.908	3.908	3.906	3.905	3.905	3.905
Evenness	0.961	0.958	0.958	0.956	0.955	0.955	0.955
Brillouin	3.862	3.859	3.858	3.856	3.852	3.853	3.856
Menhinick	0.945	0.952	0.957	0.963	0.998	0.982	0.955
Margalef	6.363	6.374	6.382	6.392	6.450	6.425	6.379
Fisher-alpha	8.919	8.946	8.964	8.987	9.124	9.064	8.958

Source: Authors' elaboration on ELSTAT and Eurostat data.

(2000–2007), the prolonged recession and initial recovery (2008–2015) and the post-crisis period including the COVID-19 pandemic (2016–2022). Together, these tables reveal persistent and intensifying patterns of regional inequality.

3.2.1. *Economic expansion (2000–2007)*

During the expansion phase (Table 2), the arithmetic mean of income rose steadily from €77.1 to €81.2, while the median increased from €71.5 to €75.5, with inequality indicators remaining high and in some cases worsening substantially. The coefficient of variation ranged from 26.6% to 31.3%, and skewness climbed from 1.3 to 1.8, indicating an increasingly asymmetric income distribution favouring a small number of wealthy prefectures. Similarly, kurtosis peaked at 3.7 in 2006, reflecting a concentration of values around the mean with some high-income outliers. The normalized range and percentile range also widened, while dominance indicators such as the Berger–Parker Index and Margalef richness showed gradual increases. However, traditional diversity indices like Shannon and Simpson remained stable, suggesting that while top incomes pulled away, the overall configuration of regional income remained structurally rigid.

3.2.2. *Economic recession (2008–2015)*

The global financial crisis severely impacted income levels across Greek prefectures (Table 3). The mean dropped from €81.5 in 2008 to €60.1 in 2015, and the median fell from €77.0 to €54.5. The coefficient of variation remained high (above 30%), confirming continued disparities, while kurtosis increased further to 4.6 in 2015, indicating a more extreme polarization in affluent and disadvantaged districts. Skewness values hovered around 1.8–1.9, and diversity measures such as the Shannon Index remained relatively unchanged, reinforcing the notion that regional inequality was embedded and largely unresponsive to macroeconomic shocks. Meanwhile, the Berger–Parker Index rose incrementally to 0.043, further confirming the growing dominance of a few high-income prefectures. Interestingly, evenness indices stayed high (>0.96), though this likely reflected the flattening of incomes across economically stagnant regions rather than improved equity.

3.2.3. *Post-crisis recovery and COVID-19 (2016–2022)*

Income levels remained stagnant during the post-crisis recovery (Table 4), with mean values hovering between €56.8 and €57.0. The COVID-19 shock further widened inequality: the coefficient of variation rose to a peak of 35.3% in 2020, while kurtosis jumped to 6.4 in 2021 – the highest across the entire series – underscoring the deepening divide between dominant urban economies and lagging rural areas. The normalized range and percentile range also hit their peaks, indicating greater dispersion across the income distribution. While the Shannon Index experienced a slight decline (to nearly 3.9), the Berger–Parker Index rose to 0.046 by 2020–2022 – its highest recorded level – signalling increased concentration of income in a few regions. Though evenness remained formally high (~0.955), it masked a structural stagnation in lower-income areas rather than genuine convergence.

3.2.4. *Overall assessment*

Taken together, the indicators reported in Tables 2–4 suggest a clear narrative: Greece's regional income inequality has remained structurally persistent through expansion, recession and crisis. Income diversity remained statistically stable, yet dominance effects increased, and polarization intensified, especially during economic downturns. These findings imply that short-term macroeconomic shifts have had limited capacity to alter the entrenched spatial inequalities in the country's economic geography.

3.3. ***Multidimensional analysis***

To further investigate the temporal and spatial dynamics of per-capita income across Greek prefectures, we employed hierarchical clustering using Ward's minimum variance method. The resulting dendrogram (Figure 1) illustrates

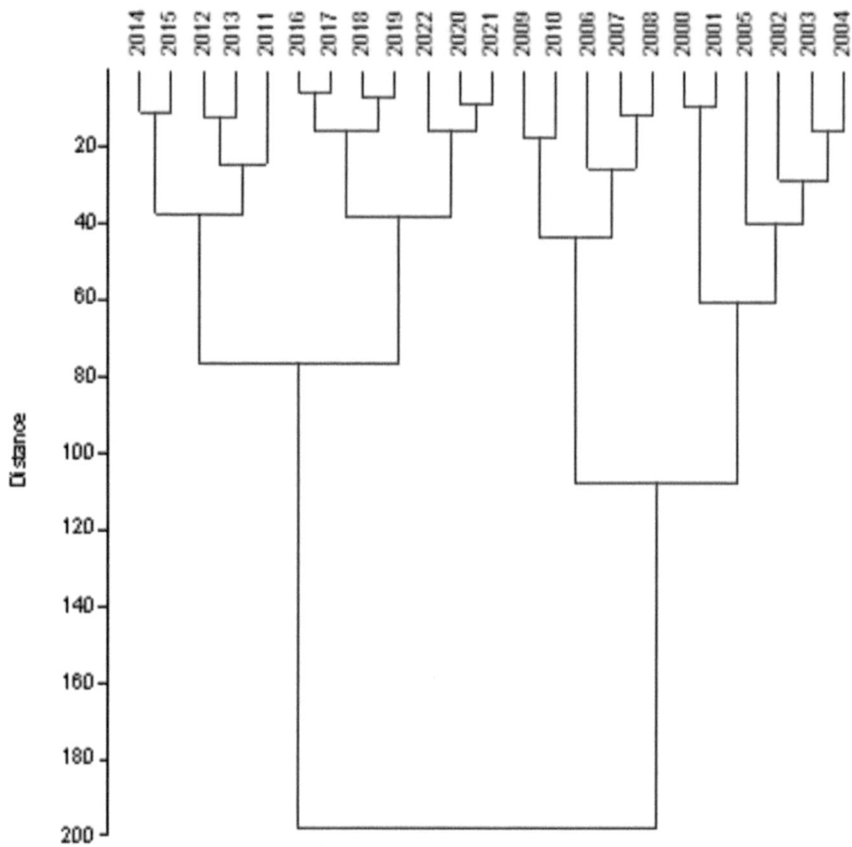

Figure 1. Hierarchical clustering of standardized per-capita income in Greece (2000–2022) by year.

Source: Authors' elaboration on ELSTAT and Eurostat data.

the standardized dissimilarities among observations from 2000 to 2022, based on the full set of diversity and distributional indicators discussed earlier. Clustering identified three distinct temporal clusters:

1. 2000–2003: This initial phase reflects a relatively stable and spatially cohesive income structure, preceding the strong expansionary effects observed in mid-decade.
2. 2004–2008: Corresponding with the height of Greece's pre-crisis growth, this period demonstrates increasing regional differentiation, particularly among urbanized and tourism-driven areas.

3. 2009–2012: Marking the early years of the economic recession, this cluster displays a sharp structural divergence from prior years, underscoring the recession's spatially uneven impact on income.

These temporal divisions highlight how the expansion and contraction phases of the economic cycle produced distinct spatial responses, reshaping regional income hierarchies. Interestingly, even the expansionary wave (2000–2008) is split into two sub-periods, suggesting intra-cycle shifts in how growth was geographically distributed. On the spatial side, clustering of prefectures based on their income profiles revealed two major regional groups, separated at approximately the 50% linkage threshold. The first group – positioned on the left branch of the dendrogram – includes urban, peri-urban, coastal and accessible interior prefectures. These are characterized by diversified economic structures, dynamic tourism sectors and relatively high resilience to economic shocks. Key members include:

- major metropolitan areas: Attica, Thessaloniki, Patra, Iraklio;
- urbanized and semi-urban prefectures: Larisa, Korinthia, Chania, Rethimno;
- high-profile tourism zones: Kyklades, Zakynthos, Kerkyra, Dodekanisos, Lasithi;
- productive rural-agricultural areas: Voiotia, Florina, Kozani, Argolida, Arkadia.

The second group, found on the right branch, encompasses remote, mountainous or otherwise disadvantaged prefectures such as Evritania, Grevena, Arta, Kastoria and Evros. These areas are marked by (i) limited economic diversification, (ii) lower accessibility and connectivity, (iii) smaller urban centres and depopulation trends. The clear separation between these clusters underscores the persistence of Greece's spatial dualisms: namely, the urban–rural and coastal–inland gradients. These structural divides appear resilient to short-term macroeconomic fluctuations, reinforcing prior findings in the literature on the path-dependency of regional inequality in Southern Europe. Figure 2 effectively visualizes these dynamics, validating the spatial stickiness of economic advantage and emphasizing how geography – rather than economic cycles alone – plays a primary role in shaping long-term regional disparities.

3.3.1. *Results of an exploratory factor analysis*

PCA summarizes recent changes in the economic geography of Greek prefectures based on per-capita income. Two principal components were extracted with eigenvalue > 1 accounting respectively for 49.4% and 20.8% of the total matrix variance (Figure 2). Component 1, according to the loadings plot (Figure 2a), strongly reflects statistical asymmetry and dispersion in the spatial distribution of income. It is positively associated with indicators such as skewness, kurtosis, and the Coefficient of variation (Cv), which highlight unequal income structures and regional polarization. This axis also correlates with variables capturing urban dominance, separating regions like Athens and Thessaloniki from the broader distribution. In contrast, lower Component 1

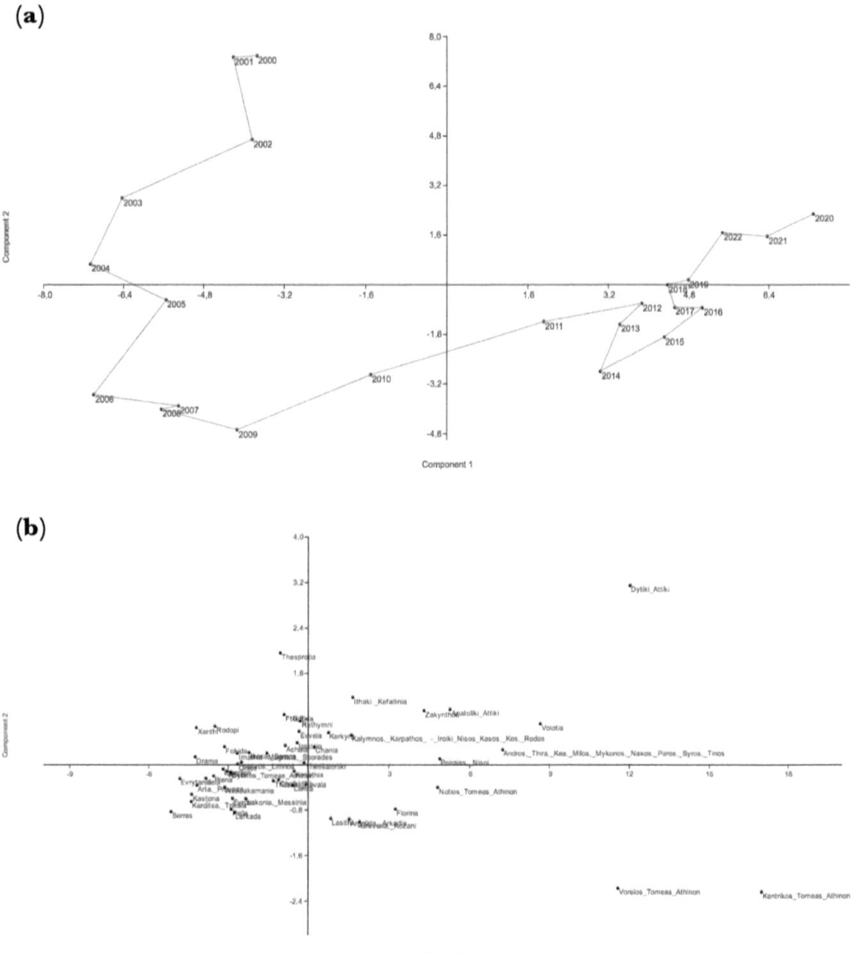

Figure 2. Principal Component Analysis of per-capita income in Greek prefectures (component 1 vs 2); (a): loadings; (b): scores.

Source: Authors' elaboration on ELSTAT and Eurostat data.

loadings are linked with indicators representing centrality and relative balance, particularly in medium-sized or secondary cities. Component 2 differentiates the structure of regional income disparities – specifically how regions deviate around the mean or the national average. High values on this axis correspond to broader percentile gaps (e.g. normalized ranges and interquartile spread), while low values emphasize the gap between Attica and the most disadvantaged prefectures, suggesting a more localized (rather than generalized) form of disparity. This component thus emphasizes the Athens-periphery divide, especially relevant during recession years. As shown in the scores plot (Figure 2b), the temporal distribution of years across the two components reveals meaningful chronological patterns.

Years associated with economic expansion (2000–2008) are located mostly on the right-hand side of Component 1, consistent with relatively lower income asymmetry and more balanced growth. The recession years (2009–2012) shift sharply leftwards along Component 1 and downwards on Component 2, reflecting a transition towards more polarized, skewed income distributions and amplified disparities between core and peripheral regions. The post-crisis and pandemic years (2013–2022) cluster separately, occupying a distinct zone in the bottom-left quadrant. This reflects structural stagnation, with persistent dispersion but without significant convergence (Proietti 2005) – indicating the emergence of a new, uneven equilibrium in Greece's regional income structure. Taken together, the PCA highlights the non-linear evolution of income geography in Greece. Economic expansion led to a moderate reduction in spatial disparities, especially across intermediate regions anchored by cities like Thessaloniki and Iraklio. Conversely, the recession phase generated greater fragmentation and asymmetry, reinforcing territorial divides and elevating the role of Athens as an outlier both in wealth and trajectory.

4. Discussion

The evolution of regional inequalities over the past two decades in Greece is not a simple story of convergence or divergence, but one of structural reconfiguration driven by macroeconomic shocks. Our findings indicate that crisis' periods do not merely alter the level of income disparity but fundamentally reshape its geography, exposing deep-seated structural weaknesses in regional economies related to industrial structures, demographics, and the spatial arrangement of economic activity (D'Uva and De Siano, 2010; Storper 2011; Lee 2012).

Our analysis of three distinct economic phases reveals a non-linear and asymmetric process. The pre-crisis expansion (2000–2007), while fostering a modest narrowing of the income gap, failed to address the underlying structural imbalances. The subsequent recessionary period (2008–2015) was not a great equalizer; rather, it triggered a significant fragmentation of the regional landscape. The widening income dispersion and increased skewness observed in our results point to a fragmented economic landscape, where business downturns created new asymmetries and pushed more regions into positions of economic distress (Taulbut and Robinson 2014). The post-crisis recovery has been incomplete, failing to reverse this divergence and instead cementing a 'new normal' characterized by persistently high variability and an unstable equilibrium (Prodromidis 2014).

Crucially, the impact of the crisis was filtered through the unique geographic and sectoral composition of Greece's regions. Our analysis challenges earlier notions that recessions foster automatic convergence (Rodríguez-Pose and Fratesi 2007) by revealing a more nuanced reality. Metropolitan Athens leveraged its concentration of high-value sectors like real estate and financial services to not only weather the storm but reinforce its economic primacy (Koutsampelas and Tsakloglou 2013). In contrast, other mid-sized urban economies, such as Thessaloniki and Iraklio, proved more vulnerable due to industrial fragmentation and reliance on external funding (e.g. Kemeny and Storper 2014). This supports the idea that, while economic diversity is a key determinant of resilience (Essletzbichler 2007), it is insufficient without structural quality and institutional capacity (Monastiriotis 2009, 2011). Perhaps most surprisingly, some of the most economically disadvantaged, agriculture-dependent, rural districts demonstrated a notable resilience, suggesting their economies are less sensitive to the shocks that reverberate through integrated urban centres (Agostino and Scarlato 2013).

These divergent regional trajectories are not random; they are rooted in Greece's enduring structural divides – most notably the gradients between coastal and inland areas and between urban and rural settlements (Kroll and Kabisch 2011) – and shaped by place-specific institutional capacity and local agency (Bristow and Healy 2014). These patterns resonate with broader European trends (King et al. 2000; Pellegrini 2002; Felice 2010) and post-crisis polarization (Benediktsson and Karlsdóttir 2011), but they are particularly entrenched in Greece, where the spatial arrangement of settlements and land use has long reflected and reinforced structural inequalities (Salvati and Carlucci 2014). The results of multivariate analysis confirm that these long-standing geographical and settlement structures are the primary factors that mediate the impact of economic cycles (Dunford 2008). This aligns with established perspectives on the role of spatial autocorrelation and local context in regional economic behaviours (Anselin 2001). The recession did not erase the map of inequality but rather redrew it more starkly along these pre-existing lines, underscoring that a region's location and functional role within the national settlement system are decisive in shaping its economic destiny.

The clear implication of these findings is the inadequacy of uniform, 'one-size-fits-all' national policies for regional development. To foster genuine and sustainable convergence, policy should become fundamentally place-based. Strategies must be tailored to address the specific vulnerabilities and leverage the inherent strengths of Greece's diverse regional economies – from its dominant metropolitan core to its resilient rural peripheries. Longitudinal analyses such as this one can help inform these crisis-responsive spatial planning frameworks by identifying the early-warning indicators of regional vulnerability (Zuindeau 2006).

5. Concluding Remarks

This study concludes that macroeconomic shocks do not simply alter the level of regional inequality in Greece, but fundamentally reconfigure its geography, deepening pre-existing structural divides. Over the 23-year period, our analysis reveals that the financial crisis, in particular, did not foster a simple convergence between affluent and disadvantaged regions. Instead, it generated greater fragmentation and asymmetry, a process that reinforced the economic primacy of metropolitan Athens while exposing the unique vulnerabilities of mid-sized urban economies and the surprising resilience of some agriculture-dependent rural districts. The primary contribution of this research is its detailed, long-term analysis that challenges simplistic convergence theories and demonstrates the 'spatial stickiness' of economic advantages in Greece. Our findings move beyond documenting the persistence of inequality to explaining how it evolves, highlighting a 'new normal' in the post-crisis era characterized by high variability and incomplete regional convergence. The path forward requires moving beyond managing crises to proactively building resilience from the ground up, acknowledging that a region's path to prosperity is inextricably linked to its geographical location.

References

Anselin, Luc. 2001. "Spatial Effects in Econometric Practice in Environmental and Resource Economics." *American Journal of Agricultural Economics* 83 (3): 705–710. http://www.jstor.org/stable/1245103.

Arbia, Giuseppe, and Jean HP Paelinck. 2003. "Economic convergence or divergence? Modeling the interregional dynamics of EU regions, 1985–1999." *Journal of Geographical Systems* 5.3: 291–314. https://doi.org/10.1007/s10109-003-0114-z.

Barro Robert J., and Xavier Sala-i-Martin. 2003. "*Economic Growth*" The MIT Press, edition 2, volume 1, number 0262025531, December.

Benediktsson, Karl, and Anna Karlsdóttir. 2011. "Iceland: Crisis and Regional Development – Thanks for All the Fish?" *European Urban and Regional Studies.*18(2): 228–235.

Bristow, Gillian, and Adrian Healy. 2013. "Regional Resilience: An Agency Perspective." *Regional Studies* 48 (5): 923–935. doi:10.1080/00343404.2013.854879.

Chapple, Karen, and William T. Lester. 2010. "The Resilient Regional Labour Market? The US Case." *Cambridge Journal of Regions, Economy and Society* 3 (1): 85–104.

Cho, Seong-Hoon, Seung G. Kim, Roland K. Roberts, Roberts, Dayton M. Lambert, and Kim Taeyoung. 2014. "Effects of Land-Related Policies on Land Development during a Real Estate Boom and a Recession." *Growth and Change.* https://doi.org/10.1111/grow.12071.

Christopoulos, Dimitris. 2004. "The Relationship between Output and Unemployment: Evidence from Greek Regions." *Papers in Regional Science* 83 (3): 611–620.

Cliff, Andrew D., and Keith J. Ord. 1981. *Spatial Processes, Models and Applications.* London: Pion.

Coppi, Renato, Sergio Bolasco, and Consiglio Nazionale delle Ricerche. 1989. *Multiway Data Analysis.* Amsterdam; North-Holland Publish Co.

Cracolici, Maria F., Miranda Cuffaro, and Peter Nijkamp. 2007. "Geographical Distribution of Unemployment: An Analysis of Provincial Differences in Italy." *Growth and Change* 38 (4): 649–670. https://doi.org/10.1111/j.1468-2257.2007.00391.x.

D'Agostino, Giorgio, and Margherita Scarlato. 2013. "Innovation, Socio-Institutional Conditions and Economic Growth in the Italian Regions." *Regional Studies* 49 (9): 1514–1534. doi:10.1080/00343404.2013.838000.

D'Uva, Marcella, and Rita De Siano. 2010. "Time Series Approaches to Italian Regional Convergence." *Applied Economics* 43 (29): 4549–4559. doi:10.1080/00036846.2010.491473.

Davies, Sara. 2011. Regional resilience in the 2008–2010 downturn: comparative evidence from European countries, *Cambridge Journal of Regions, Economy and Society*, Volume 4, Issue 3, November, Pages 369–382, https://doi.org/10.1093/cjres/rsr019.

Dixon, Robert, David Shepherd, and James Thomson. 2001. "Regional Unemployment Disparities in Australia." *Regional Studies* 35 (2): 93–102. doi:10.1080/00343400120033089.

Dunford, Michael. 2008. "The North-South Divide after the Three Italies (Internally Differentiated): Analyzing Regional and Industrial Trajectories." *Annales de géographie*, 2008/6 No 664, 2008. p. 85–114. https://doi.org/10.3917/ag.664.0085.

Duran, Engin H. 2014. "Short-Run Dynamics of Income Disparities and Regional Cycle Synchronization in the U.S." *Growth and Change* 45: 292–332. https://doi.org/10.1111/grow.12040.

ESPON (The European Spatial Planning Observation Network). 2014. *Territorial Monitoring Report: Progress towards the Territorial Agenda of the European Union 2020.* Luxembourg: ESPON Coordination Unit. https://archive.espon.eu/topics-policy/publications/territorial-observations/territorial-monitoring-report.

Essletzbichler, Jürgen. 2007. "Diversity, Stability and Regional Growth in the United States (1975–2002)." In *Applied Evolutionary Economics and Economic Geography*, edited by Koen Frenken, 203–229. Cheltenham: Edward Elgar.

Ezcurra, Roberto, Pedro Pascual, and Manuel Rapún. 2007. "The Dynamics of Regional Disparities in Central and Eastern Europe during Transition." *European Planning Studies* 15 (10): 1397–1421. doi:10.1080/09654310701550850.

Felice, Emanuele. 2010. "Regional Development: Reviewing the Italian Mosaic." *Journal of Modern Italian Studies* 15 (1): 64–80. doi:10.1080/13545710903465556.

Fingleton, Bernard, Harry Garretsen, and Ron Martin. 2012. "Recessionary Shocks and Regional Employment: Evidence on the Resilience of U.K. Regions." *Journal of Regional Science* 52 (1): 109–133.

Fratesi, Ugo, and Marco Percoco. 2013. "Selective Migration, Regional Growth and Convergence: Evidence from Italy." *Regional Studies* 48 (10): 1650–1668. doi:10.1080/00343404.2013.843162.

Frenken, Koen, Frank Van Oort, and Thijs Verburg. 2007. "Related Variety, Unrelated Variety and Regional Economic Growth." *Regional Studies* 41 (5): 685–697. doi:10.1080/00343400601120296.

Glaeser, Edward L., Giacomo A. M. Ponzetto, and Kristina Tobio. 2012. "Cities, Skills and Regional Change." *Regional Studies* 48 (1): 7–43. doi:10.1080/00343404.2012.674637.

Henning, Martin, Erik Stam, and Rik Wenting. 2013. "Path Dependence Research in Regional Economic Development: Cacophony or Knowledge Accumulation?" *Regional Studies* 47 (8): 1348–1362. doi:10.1080/00343404.2012.750422.

Holm, Jacob R., and Christian R. Østergaard. 2013. "Regional Employment Growth, Shocks and Regional Industrial Resilience: A Quantitative Analysis of the Danish ICT Sector." *Regional Studies* 49 (1): 95–112. doi:10.1080/00343404.2013.787159.

Kemeny, Thomas, and Michael Storper. 2014. "Is Specialization Good for Regional Economic Development?" *Regional Studies.* 49 (6): 1003–1018. doi:10.1080/00343404.2014.899691.

King, Russell, Paola De Mas, and Jan Mansvelt Beck, eds. 2000. *Geography, Environment and Development in the Mediterranean.* Brighton: Sussex Academic Press. https://hdl.handle.net/11245/1.207817.

Koutsampelas, Christos and Panos Tsakloglou. 2013. The distribution of full income in Greece. *International Journal of Social Economics* 8 March 2013; 40 (4): 311–330. https://doi.org/10.1108/03068291311305008.

Kroll, Franziska and Nadja Kabisch. 2011. "The Relation of Diverging Urban Growth Processes and Demographic Change along an Urban-Rural Gradient." *Population, Space and Place* 18 (3): 260–276. https://doi.org/10.1002/psp.653.

Lee, Neil. 2012. "Grim down South? The Determinants of Unemployment Increases in British Cities in the 2008–2009 Recession." *Regional Studies.* 48 (11): 1761–1778. doi:10.1080/00343404.2012.709609.

Leontidou, Lila and Emmanuel Marmaras. 2001. "From Tourists to Migrants: Residential Tourism and 'Littoralisation'." In *Mediterranean Tourism: Facets of Socioeconomic Development and Cultural Change.* edited by Yorgos Apostolopoulos, Lila Leontidou, Philippos Loukissas. Routledge.

Liao, Felix H. F., and Yehua Dennis Wei. 2012. "Dynamics, Space, and Regional Inequality in Provincial China: A Case Study of Guangdong Province." *Applied Geography* 35 (1–2): 71–83. https://doi.org/10.1016/j.apgeog.2012.05.003.

Liberati, Paolo. 2015. "The World Distribution of Income and Its Inequality, 1970–2009." *Review of Income and Wealth* 61 (2): 248–273. https://doi.org/10.1111/roiw.12088.

Martin, Ron. 2012. Regional economic resilience, hysteresis and recessionary shocks, *Journal of Economic Geography*, 12 (1): 1–32. https://doi.org/10.1093/jeg/lbr019.

Martin, Ron, and Peter Sunley. 2014. "Towards a Developmental Turn in Evolutionary Economic Geography?" *Regional Studies* 49 (5): 712–732. doi:10.1080/00343404.2014.899431.

Monastiriotis, Vassilis. 2009. "Examining the Consistency of Spatial Association Patterns across Socio-Economic Indicators: An Application to the Greek Regions." *Empirical Economics* 37 (1): 25–49. https://doi.org/10.1007/s00181-008-0221-2.

———. 2011. Making geographical sense of the Greek austerity measures: compositional effects and long-run implications, *Cambridge Journal of Regions, Economy and Society*, 4 (3): 323–337. https://doi.org/10.1093/cjres/rsr026.

Patacchini, Eleonora, and Patricia Rice. 2007. "Geography and Economic Performance: Exploratory Spatial Data Analysis for Great Britain." *Regional Studies* 41 (4): 489–508. doi:10.1080/00343400600928384.

Patacchini, Eleonora. 2008. "Local Analysis of Economic Disparities in Italy: A Spatial Statistics Approach." *Statistical Methods & Applications* 17 (1): 85–112. https://doi.org/10.1007/s10260-007-0054-8.

Pellegrini, Guido. 2002. "Proximity, Polarization and Local Labour Market Performances." *Networks and Spatial Economics* 2 (2): 151–174. https://iris.uniroma1.it/handle/11573/13935.

Pendall, Rolf, Kathryn A. Foster, and Margaret Cowell. 2010. "Resilience and Regions: Building Understanding of the Metaphor." *Cambridge Journal of Regions, Economy and Society* 3 (1): 71–84. https://doi.org/10.1093/cjres/rsp028.

Prodromidis, Prodromos. 2014. *The Evolution of Personal Incomes Across Greece 2001–2008: Seeing the Economy Through a Micro-regional Lens and Developing a Method for Identifying Policy Areas.* Athens: Stamoulis Publications.

Proietti, Tommaso. 2005. "Convergence in Italian Regional Per-Capita GDP." *Applied Economics* 37 (5): 497–506. doi:10.1080/0003684042000318173.

Rodríguez-Pose, Andrés, and Ugo Fratesi. 2007. "Regional Business Cycles and the Emergence of Sheltered Economies in the Southern Periphery of Europe." *Growth and Change* 38 (4): 621–648. https://doi.org/10.1111/j.1468-2257.2007.00390.x.

Salvati, Luca, Marco Zitti, and Adele Sateriano. 2013. "Changes in City Vertical Profile as an Indicator of Sprawl: Evidence from a Mediterranean Urban Region." *Habitat International* 38: 119–125. https://doi.org/10.1016/j.habitatint.2012.05.006.

Salvati, Luca, Giuseppe Venanzoni, and Margherita Carlucci. 2014. "Urban Growth and Polycentric Settlements: An Exploratory Analysis of Socioeconomic and Demographic Indicators in Emilia Romagna, Italy." *Statistica Applicata* 26 (3): 167–178.

Salvati, Luca and Margherita Carlucci. 2014. "Exploratory Spatial Data Analysis of Land-Use Changes along the Urban-Rural Gradient: The Impact of City's Form." *SAGE Open* 4 (4): 1–13. https://doi.org/10.1177/2158244014561199.

Sicular, Terry, Yue Ximing, Björn Gustafsson, and Li Shi. 2007. "The Urban–Rural Income Gap and Inequality in China." *Review of Income and Wealth* 53: 93–126. https://doi.org/10.1111/j.1475-4991.2007.00219.x.

Storper, Michael. 2011. "Why Do Regions Develop and Change? The Challenge for Geography and Economics." *Journal of Economic Geography* 11 (2): 333–346. https://doi.org/10.1093/jeg/lbq033.

Taulbut, Martin, and Mark Robinson. 2014. "The Chance to Work in Britain: Matching Unemployed People to Vacancies in Good Times and Bad." *Regional Studies* 49 (12): 2070–2086. doi:10.1080/00343404.2014.893058.

Terrasi, Massimo. 1999. "Convergence and Divergence across Italian Regions." *Annals of Regional Science* 33: 491–510. https://doi.org/10.1007/s001680050117.

Tsakloglou, Panos. 1990. "Aspects of Poverty in Greece." *Review of Income and Wealth* 36: 381–402. https://doi.org/10.1111/j.1475-4991.1990.tb00321.x.

Van Well, Lisa. 2012. "Conceptualizing the Logics of Territorial Cohesion." *European Planning Studies* 20 (9): 1549–1567. doi:10.1080/09654313.2012.708021.

Zuindeau, Bertrand. 2006. "Spatial Approach to Sustainable Development: Challenges of Equity and Efficacy." *Regional Studies* 40 (5): 459–470. doi:10.1080/00343400600757437.

Chapter 3

REGİONAL DEVELOPMENT, OPTİMAL SUSTAİNABİLİTY MANAGEMENT AND THE SUBJECTİVE ASSESSMENT OF LİFE QUALİTY İN A PRE-CRİSİS ECONOMİC LANDSCAPE

Vasiliki Benaki, Adele Sateriano, Luigi Fici, Marco Maialetti, Donato Scarpitta, Maurizio Masi, Ioannis Konaxis and Luca Salvati

1. Introduction

Composite indexes are key tools measuring the integrated performance of multifaceted (e.g. social, environment, economic) processes. An issue underlying indication theory is how comprehensive and reliable is the information held by such complex tools. Composite indexes can be biased tools if the assumptions underlying the adopted 'indication theory' are poorly (or partly) specified (Salvati and Zitti 2009). Use of elementary indicators from official statistics is a measurement challenge in social science, since it may easily point out both economic growth and processes of technological change, contributing to prioritize specific developmental policies (Barro and Sala-i-Martin 2003). Composite indexes aim at assessing truly kaleidoscopic notions by reducing (and possibly, re-organizing) the quantitative information stemming from a full set of elementary indicators, eventually synthesized using appropriate statistical techniques (OECD/European Union/EC-JRC 2008).

In this perspective, instead of a one-by-one scrutiny of a wide system of elementary indicators, composite indexes are usually easily interpreted and can include relevant information appropriately organized and summarized for basic users (e.g. Zambon et al. 2017). This facilitates the dissemination of research outcomes to stakeholders. However, composite indexes are potentially more unstable than individual indicators from the statistical perspective

(Salvati et al. 2013). Consequently, there is a risk of reducing policy implications stemming from composite indexes to rather simplistic conclusions in order to avoid the negative impact of misleading (regional/local) information (Booysen 2002). This aspects adds to the intrinsic (positive and normative) challenge of identifying proper development solutions to already complex economic paths of growth and change (Freudenberg 2003).

Based on these premises, setting up a list of (qualitative and quantitative) criteria applied to elementary indicators when developing composite indexes is mandatory (Michalos et al. 2010; OECD/European Union/EC-JRC 2008). Being relevant to specific target audiences, the individual development of any composite index should be anchored in a well-established theory assumed to be easily understandable and comparable across standard classifications, official criteria, as well as temporal and spatial dimensions (Cecchini et al. 2019). An additional criterion is integrity, which secures the final index from political bias, being objective, reliable and verifiable (Saisana et al. 2005). Once consolidated, composite indexes should be subjected to a democratic, open and transparent consultation process (or peer-review) in order to question their validity and to enhance circulation of new ideas and practical solutions to complex monitoring issues (Ciommi et al. 2018). Last but not least, input data should be easy to obtain, e.g. from official sources, in order to enable periodic and timely update (Saisana and Tarantola 2002).

Subjective choices (and the related indicators) are needed to prevent the achievement of some goals at the expense of others (Lamonica et al. 2019). For instance, use of traditional approaches/indicators that are easy to compute, collect and/or update, should not discourage further investigation (Funtowicz and Ravetz 1990). The operational procedure of any index composition varies according with the underlying dimensions/variables/indicators being aggregated, where issues such as combining individual and separated information domains, and defining their weighting, are substantial to the success of the statistical procedure (Salvati and Carlucci 2015). When compiling a systemic procedure which serves policy making in a given dimension, variables' selection and indicators' specification are the first operational steps having an intrinsic importance in any summary index (Salvati et al. 2019). The subsequent stages in the compositional process include (i) indicators' collection and data manipulation through simple (usually descriptive) statistical tools and procedures, (ii) methodologies allowing composition of the individual indicators into a unique index summarizing the collected information in the original database, and (iii) an explicit analysis evaluating the reliability of the final index (Ciommi et al. 2019). The focal point of this study is the meaningful aggregation of separate research components by considering different methodologies and valuating their suitability when constructing a composite index of QoL (Bajocco et al. 2016). We assumed a perspective grounded on the 'subjective' dimension of well-being as an original research issue (both positive and normative) for this study.

1.1. *QoL as a key research issue in regional science*

QoL is a concept that implies the multidimensional evaluation of the overall quality of a person's whole life (e.g. Zambon et al. 2017; Diener 2009). Since the seminal work by Andrews (1976) and Campbell et al. (1976), scholars have considered QoL as a notion composed of several, different domains that should be carefully defined, specified and measured. While being particularly appropriate in the construction of any QoL index, the operational specification of some notions such as 'health' or 'relational networks' to designate defined parts of the life experience, is a challenging task (Pappa et al. 2009). This operation can be complemented with an easier specification – possibly based on objective measures – of well-known socio-economic dimensions such as 'accommodation conditions' and 'accessible durable goods' – that also represent relevant domains contributing to QoL (Greyling and Tregenna 2017).

Despite a partial consensus on QoL definition and assessment techniques exists in advanced economies, some divergent opinions remain on the number and characteristics of the composing domains and their operational measurement (Bericat et al. 2019). A relatively large diversity of composing domains was proposed, encompassing a total of 351 terms that were supposed to aliment different QoL definitions – both theoretically oriented and empirically driven (Ivaldi et al. 2014). These definitions were basically objective (e.g. based on indicators specifically selected by the operator) or subjective (e.g. selected and ranked for importance by a representative panel of respondents) (Efklides et al. 2003). For the latter operation, a number of surveys was implemented with the aim at assessing perceived QoL (Cummins 1997). Finally, governments and public authorities, including policy-makers, have frequently developed Quality of Life Indicators (QLI) based on both objective and subjective approaches (European Commission: Eurostat 2015).

Based on these premises, this study encompasses various stages requested for the construction of a composite index and points out some technical issues arising when indicators assessing QoL are routinely implemented. The study was carried out in Greece using freely available statistical data reporting distinctive thematic dimensions derived from Hellenic Statistical Authority (ELSTAT) at the NUTS-2 administrative level (Gavalas et al. 2014). Indicators refers to an observation time interval reflecting a pre-crisis economy under mild growth – the most frequent condition in the history of the country after World War II. This condition can be roughly generalized to other countries in Mediterranean Europe with similar post-war socio-economic dynamics (Benyaklef 1997). More recent data, mainly covering the crisis time (e.g. since 2007), are reflective of particularly harsh socio-economic conditions that were partly recovered over time, and that made QoL estimations especially dependent on short-term economic disturbances (Monastiriotis 2011).

The approach adopted in this study is assumed to be flexible enough to guarantee full substitutability and continuous update of the selected indicators with an enriched information set, benefiting from future technical (and technological) improvements. In this perspective, the empirical results illustrated here can be taken as the outcome of a preliminary and basic exercise that may shed light on the intrinsic mechanisms determining (or contributing to) regional disparities in a marginal (and mostly divided) country in Southern Europe. The exercise can be improved with additional information and can be generalized to similar socio-economic contexts in Mediterranean Europe and, possibly, to other marginal contexts in the old continent. Implementation of an integrated methodology finally highlights the importance of composite indexes (i) estimating socio-economic disparities and (ii) informing specific developmental policies for disadvantaged regions in advanced countries (Koutsampelas and Tsakloglou 2013; Christopoulos 2004).

2. Study Area

This study investigates Greece, a Mediterranean region extending about 132 thousands km^2 and divided in 13 administrative regions and more than 50 prefectures (Figure 1). Administrative regions in Greece correspond to the Nuts-2 level of European Nomenclature of Territorial Statistical Units (Pili et al. 2017). Based on data availability, administrative regions are regarded a reliable unit allowing a comparative scrutiny of QoL as a subjective (i.e. based on individual perceptions) notion measured through official surveys, i.e. assuring reliable and precise information based on dedicated sampling designs and appropriate survey efforts. Greek settlements are heavily divided, since more than half population (more than 10 million inhabitants) took residence in metropolitan Athens or Thessaloniki (respectively, corresponding to Attica and Central Macedonia regions) (Maloutas 2007). Both cities expanded largely since World War II because of various reasons, and took the lead of the national hierarchy overpassing largely the remaining cities – expanding broadly only in the 1970s and the 1980s (Chorianopoulos et al. 2014). Urban population stabilized in recent decades (Zambon et al. 2017), while internal districts were increasingly exposed to depopulation and economic shrinkage (Figure 2) (Kasimis 2008). Especially in the Aegean regions (Crete, Cyclades,

(a)

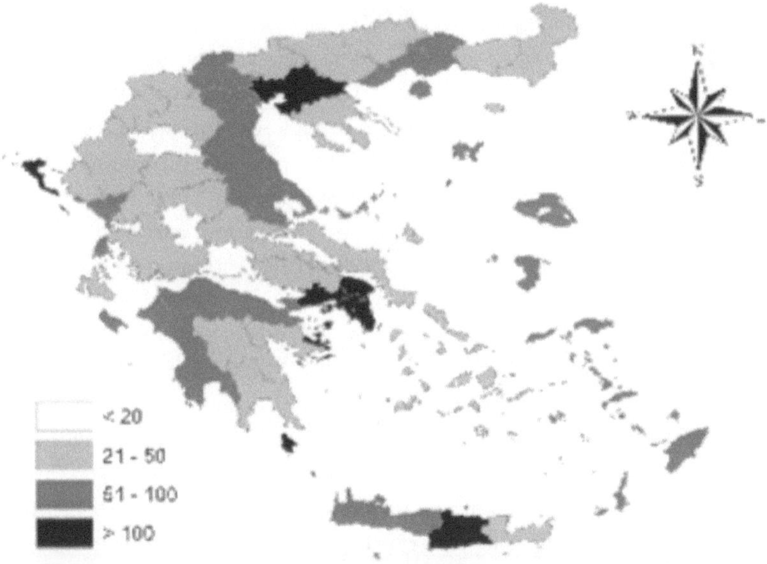

(b)

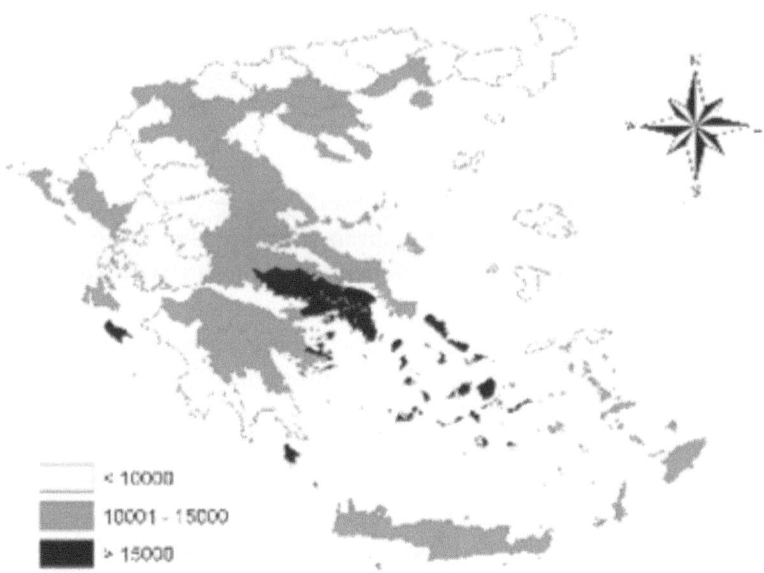

Figure 1. Selected indicators delineating key socio-economic aspects of Greek regions (a: population density, inhabitants/km²; b: per-capita income, Euros.

Source: Authors' elaboration on official statistics released by Hellenic Statistical Authority, ELSTAT).

Dodecanese), tourism-specialized districts with a most dynamic (rural) economy, started expanding since the 1990s (Carlucci et al. 2017).

Recession since 2007 has negatively (and significantly) affected any aspect of individual and collective life in both urban and rural locations (Tragaki and Bagavos 2019). Assuming QoL as a function of socio-demographic conditions, this study has investigated pre-crisis (2003–2004) dynamics as a representative time of stable economic conditions and mild social development. These conditions reflect a dominant (long-term) development path in Greece, and can be generalized to other countries in Mediterranean Europe (Di Feliciantonio and Salvati 2014). Since the late-2000s, economic crisis has drastically altered such conditions bringing to unexpected, and often unwanted, transformations in Greece and, similarly, in other Southern European countries (Matsaganis and Leventi 2012). These transformations have determined unstable social contexts that make the estimation of QoL more volatile and basically dependent on the specific stage of the economic cycle. The empirical findings of our exercise are thus reflective of such conditions (namely, economic stability and mild social development) and can be generalized by controlling for the existence of such conditions in other study areas and socio-demographic backgrounds (Rodríguez-Pose and Fratesi 2007).

3. Logical Framework

Theory is increasingly requested to relate an interpretative model of the observed (e.g. natural/social/economic) system with the causality associated with productive contexts (Henning et al. 2013). The model should fit the specific objective of any user, being an appropriate tool that reflects the objectives motivating this output. The resulting framework implies the recognition of the multidimensional nature of a given phenomenon, justifying the full effort of specifying individual components and their (latent) relationship. According to OECD, European Union and Joint Research Centre (2008), theoretical frameworks better addressing complex issues depend on specific research challenges and issues that can be summarized into three operational dimensions: (i) structuring the framework in relation with different aspects (namely, technical, economic, social, institutional, territorial, cultural), geographical levels and/or spatial scales; (ii) addressing several disciplinary and practical perspectives identified across stakeholders and, finally, (iii) handling the degree of uncertainty in a comparative analysis of complex systems.

3.1. *Selection of key variables and relevant indicators*

Being derived from any relevant variable, elementary indicators were selected on the basis of high quality standards that meet user perspectives, needs and priorities. However, the inherent heterogeneity of such characteristics for different user groups is a multifaceted and challenging aspect that is hardly defined and quantifiable (Costa 2015). Establishing a general system of quality control for statistical products is an important issue leading countries and international organizations to introduce methodologies and approaches quantifying the overall quality of statistics and indicators produced (Saisana et al. 2005). In this perspective, five quality dimensions have been usually considered including:

a. integrity (individual characteristics that support adherence to an objective production of official statistics and elementary indicators);
b. methodological soundness (current practices related to international standards for collection/survey of specific statistics and indicators);
c. accuracy (data sources, statistical techniques and weighting systems, adequate to represent the context under investigation);
d. reliability (users' needs meeting timeliness of statistical products, their frequency, consistency and life cycle).

The above criteria are pivotal for the selection of key variables and the construction of appropriate elementary indicators that can be subsequently summarized into composite indexes (García-Bernabeu et al. 2022). However, the appropriateness of the indicators adopted for measurement of the theoretical concept under investigation, determined the overall quality of any final index. Strengths and weaknesses of the indicators selected and the related data characteristics (e.g. availability, type, source) should be clarified to assist stakeholders in assessing the overall quality of any composite index (Nardo et al. 2005).

3.2. *Multivariate analysis*

Multidimensional statistical techniques support model builders in a honest evaluation of the statistical suitability of any input dataset for identification of key variables and relevant indicators, and to provide a refined understanding and definition of the subsequent methodological steps, namely indicators' weighting and aggregation. Different analytical methods and approaches can be used, making any decision of the proper technique quantifying a given conceptual issue a mostly subjective task (Nardo et al. 2004). Three approaches, taken as the most frequently used in the technical literature, were proposed and briefly discussed as follows, based on specific statistical problems.

PCA was initially developed with the aim at deriving a general profile of the different types of indicators and the latent correspondence with each region under investigation, on the basis of objective and subjective information. A set of descriptors has been identified that maximized the informative content of the survey, with no missing data – incompatible with the application of such methodologies. These descriptors provide a prospective view of the aspects subjected to measurement in the sample survey (Tarantola et al. 2002).

In addition to PCA, we used Cronbach alpha coefficients as a statistical measure assessing the internal consistency (or reliability) of a set of scale or test items, quantifying how closely related a group of items are as a whole, and definitely indicating the extent to which they measure the same underlying construct. The value of Cronbach alpha ranges from 0 to 1, with higher values (typically above 0.7) indicating greater internal consistency. However, extremely high values (e.g. above 0.95) may suggest redundancy among items (Nunnally 1978).

Having evaluated the internal consistency of input datasets, hierarchical clustering – a statistical technique used to group data points into a hierarchy of clusters (Anderberg 1973) based on their similarity – was finally adopted with the aim at building a tree-like structure (dendrogram) which illustrates the nested grouping of data at various levels of similarity (OECD/European Union/EC-JRC 2008). Similarity between clusters was measured using distance metrics such as Euclidean distance, and linkage criteria including a particularly effective measure such as Ward's metric (Massart and Kaufman 1983).

3.3. *Data transformation, weighting and aggregation*

Algebraic transformations can be applied before data aggregation (Keeney and Raiffa 1993). Several transformations (both linear and non-linear) were introduced for this objective (Jacobs 2004). Linear transformations using z-scores, ranking (being less affected by statistical outliers) and non-linear (min–max) data re-arrangements (Saisana and Tarantola 2002) seem to be simple, broadly used procedures in this perspective (Freudenberg 2003). Subsequent to any algebraic transformation, weighting the different components of a summary index in a way that reflects their significance, statistical adequacy and timeliness, among others, is a key statistical aspect. Equal weighting is likely the simplest aggregation way. Additionally, PCA is an useful technique in this operational step when the correlation structure of variables/indicators is assumed as a preferential weighting criterion. In this perspective, Confirmatory Factor Analysis (CFA) is usually performed with the aim at examining the intimate relationship between variables. Both PCA and CFA may capture homogeneity when illustrating multidimensionality. Participatory approaches can finally contribute to a better knowledge of defined policy targets, especially measures included in National Plans, strategic documents or other formal procedures. For international comparisons, however, such references are sometimes unavailable, partially lacking or hardly achievable with standard approaches. Aggregation procedures depend on the correlation between the individual indicators and data comparability (Funtowicz and Ravetz 1990), the most popular techniques being linear or geometric (Keeney and Raiffa 1993; Pareto 2013).

3.4. *Facts and findings from an preliminary exercise*

Based on the operational framework illustrated above, a simplified exercise initially considering seven indicators was proposed here. These indicators allow a subjective QoL assessment based on aggregate information derived from official statistics at the level of administrative regions (Nuts-2). Such information include the per cent share of households reporting (i) lack of car or van, (ii) lack of telephone, (iii) lack of home computer, (iv) lack of bath (or shower) in the residence dwelling, (v) lack of hot running water in the residence dwelling, (vi) pollution and other environmental issues derived from traffic, urban congestion or industrial development in the surrounding of the residence dwelling and, finally, (vii) vandalism or crime in the settlement area. These figures were specifically derived from Eurostat-harmonized sampling surveys held by ELSTAT (Hellenic Statistical Authority) – the so called EU-SILC (Survey on Income and Living Conditions). Based on earlier studies, these indicators evaluate simplified – while different – subjective dimensions of QoL and may suggest multiple agendas for political action, from both social and economic perspectives (Papageorgiou et al. 2005). Moreover, these indicators revealed suitable to describe key subjective aspects of QoL in both urban and rural areas, exploiting the high (technical) value of official statistics (e.g. controlled surveys, fixed sampling errors, a structured network of respondents, post-survey informatics checks). From the geographical perspective, in the following parts of the chapter, we mentioned the individual Greek regions using standard transliteration rules; in sake of clarity, we reported here a complete list of Greek regions using English names (Eastern Macedonia, Central Macedonia, Western Macedonia, Thessaly, Epirus, Ionian islands, Western Greece, Central Greece, Peloponnese, Attica, Northern Aegean, Southern Aegean, Crete) and the corresponding Greek name (transliterated) in the same order (An. Makedonia, K. Makedonia, D. Makedonia, Thessalia, Ipeiros, Ionia Nisia, Dytiki Ellada, Sterea Ellada, Peloponissos, Attiki, Voreio Aigaio, Notio Aigaio, Kriti).

4. Empirical Results

The statistical distribution of the seven indicators taken as individual components of QoL in Greek regions was reported in Table 1. While some indicators, such as vandalism/crime or pollution, seem to be more associated to metropolitan regions such as Attica (Athens) or Central Macedonia (Thessaloniki), or similar areas in other parts of the country, other indicators increased evidently in rural districts. Just to mention the most relevant, this is the case for the lack of bath or shower in the residence dwelling or the lack of hot water.

4.1. *Multidimensional analysis*

PCA determines a non-parametric correlation matrix among indicators (Table 2) showing which indicators have the highest (Spearman) pair-wise coefficients. Four coefficients were found significant ($p < 0.05$) after Bonferroni's correction for multiple comparisons.

Table 3 reports the eigenvalues of the correlation matrix of the seven indicators assessing QoL in Greek regions. The first component accounted for the highest variance of the input matrix (3.6); the second component explained the highest amount of the remaining variance, and so on. The third component identified an eigenvalue close to 1 and the last four components accounted for a residual variance (13%). Based on the eigenvalue criterion adopted in this study, two components were selected for the subsequent analysis. Component 1 relates with per cent shares of households reporting lack of car (or van), telephone, home computer, bath (or shower), and hot running water in the main residence dwelling. Component 2 relates with the per cent share of households reporting pollution and other environmental issues caused by traffic or industry, as well as vandalism (or crime) in the settlement area.

Cronbach coefficient (alpha) observed for Component 1 amounted to 0.743 and demonstrated indirectly how the selected indicators measure the latent phenomenon satisfactory. Table 4 reports a basic example of Cronbach alpha delineating the intrinsic correlation with the total indicators' matrix after deleting one indicator step-by-step, as typical of reliability and redundancy analysis. The per cent share of households reporting lack in home computer has displayed the highest correlation for an individual variable. If deleted, the coefficient alpha for this variable would be as low as 0.66. If the per cent share of households reporting lack of telephone line is deleted, Cronbach alpha would increase to 0.76. Cronbach coefficient alpha for all variables was 0.65, indirectly documenting how Component 2 has slightly and negatively affected the internal consistency of the whole indicators' set.

Table 1. The distribution of each indicator in Greek regions (percentage of households reporting selected features of interest in QoL monitoring).

Region	Lack of Car or Van	Lack of Telephone	Lack of Home Computer	Lack of Bath or Shower	Lack of Hot Water	Pollution, Crime, Ecological Issues (Traffic, Industry)	Vandalism or Crime
An. Makedonia	83.0	15.7	81.6	14.0	15.8	5.9	2.2
K. Makedonia	69.7	3.0	74.1	6.3	28.1	24.9	22.8
D. Makedonia	29.3	1.2	36.0	3.1	6.8	27.8	1.5
Thessalia	68.7	4.3	79.5	12.1	31.7	10.0	1.2
Ipeiros	76.2	4.0	79.4	17.8	35.1	0.8	2.9
Ionia Nisia	37.6	4.2	67.2	6.8	46.8	1.7	0.9
Dytiki Ellada	77.8	8.1	83.9	6.3	31.6	3.6	0.6
Sterea Ellada	27.0	1.6	40.3	1.1	10.5	2.3	1.3
Peloponissos	31.1	1.8	42.1	4.0	14.2	4.6	2.3
Attiki	29.7	1.4	29.0	0.4	27.9	24.9	12.6
Voreio Aigaio	64.0	2.8	59.3	9.5	31.0	3.2	1.1
Notio Aigaio	31.9	1.2	34.1	1.3	18.4	4.9	0.9
Kriti	67.6	6.8	73.8	18.8	34.2	19.1	2.5

Source: Authors' elaboration on official statistics released by Hellenic Statistical Authority, ELSTAT.

Table 2. Correlation matrix for the selected QoL indicators (per cent share of household reporting the indicated features in total households.

Variable	Lack of Telephone	Lack of Home Computer	Lack in Bath or Shower	Lack in Hot Running Water	Pollution, Crime, other Environ. Issues	Vandalism or Crime
Lack of car or van	**0.71***	**0.92***	**0.79***	0.41	−0.15	0.08
Lack of telephone		0.69	0.57	0.11	−0.21	−0.19
Lack of home computer			**0.78***	0.56	−0.26	−0.01
No bath (or shower)				0.46	−0.16	−0.17
No hot running water					−0.19	0.08
Crime/ecological problems						0.62

Source: Authors' elaboration on official statistics released by Hellenic Statistical Authority, ELSTAT.
Note: Asterisk indicates 'significant' coefficients at $p < 0.05$ after Bonferroni's correction for multiple comparisons.

Table 3. The empirical results of Principal Component extraction.

Component	**Initial Eigenvalues**			**Extraction Sums of Squared Loadings**			**Rotation Sums of Squared Loadings**		
	Total	% Variance	Cumulative %	Total	% of Variance	Cumulative %	Total	% of Variance	Cumulative %
1	3.58	51.10	51.10	3.56	51.10	51.10	3.49	49.88	49.88
2	1.58	22.63	73.73	1.58	22.63	73.72	1.67	23.85	73.72
3	0.93	13.32	87.05						
4	0.45	6.43	93.47						
5	0.27	3.85	97.32						
6	0.13	1.89	99.22						
7	0.06	0.78	100.0						

Source: Authors' elaboration on official statistics released by Hellenic Statistical Authority, ELSTAT.
Note: See also Tables 1 and 2; the first two components selected and processed further.

Table 4. An example of Cronbach alpha analysis for Component 1.

Variable	Scale Mean if Item Deleted	Scale Variance if Item Deleted	Corrected Item-Total Correlation	Cronbach's Alpha (if Item Deleted)
Households reporting lack of car or van	157.8	3176.4	0.899	0.534
Households reporting lack of telephone	206.9	5488.3	0.661	0.756
Households reporting lack of home computer	91.2	1331.1	0.944	0.658
Households reporting no bath or shower	203.4	5147.5	0.798	0.724
Households reporting no hot running water	185.6	4948.3	0.506	0.725

Source: Authors' elaboration on official statistics released by Hellenic Statistical Authority, ELSTAT.

4.2. *Data transformation, weighting and aggregation*

In this study, per cent ranking [0–100] of individual indicators was adopted as the most pertinent transformation approach, making the identification of different levels of regional performance easier and immediate. The advantage of this approach is its simplicity and the independence from outliers (Nardo et al. 2004). Weighting was subsequently realized considering the results of PCA. The first two components illustrate, together, QoL dimensions such as accommodation and durables, whereas Component 2 loadings uniquely represent socio-environmental issues. Individual weights were derived from normalized square loadings with their sum equal to one, in order to assure comparability (Table 5).

In this study, a linear aggregation of elementary indicators was developed by summing normalized and weighted individual indicators, with weights derived from the multivariate analysis, as recalled above (Table 6). The empirical results of the PCA as far as QoL is concerned, identified two relevant dimensions in the spatial organization of the investigated regions.

4.3. *A confirmatory cluster analysis*

To link elementary indicators and the specific information stemming from the selected composite index, a hierarchical tree-clustering with average

Table 5. Indicators' weighting.

Variable	Prin. Component		Final
	1	2	Weight
Households reporting lack of car or van	0.247	0.087	0.109
Households reporting lack of telephone	0.240	−0.002	0.106
Households reporting lack of home computer	0.240	0.014	0.106
Households reporting absence of bath or shower	0.237	0.092	0.105
Households reporting absence of hot running water	0.162	−0.157	0.072
Households reporting pollution, crime, ecological problems	0.011	0.559	0.247
Households reporting vandalism or crime in the area	0.076	0.580	0.256

Source: Authors' elaboration on official statistics released by Hellenic Statistical Authority, ELSTAT.

linkage and squared Euclidean distances was adopted in this study (Figure 2). The final dendrogram identifies the greatest dissimilarity among the 13 Greek regions at a linkage distance close to 6–7, suggesting an optimal partition formed of 6 groups: (i) Sterea Ellada, Peloponnisos, Notio Aigaio, (ii) Dytiki Makedonia, (iii) Attiki, (iv) Thessalia, Ipeiros, Dytiki Ellada, Kriti, Thraki, Kentriki Makedonia, (v) Ionia Nisia and (vi) Voreio Aigaio.

A non-hierarchical cluster was also run on the same dataset using a *k*-means algorithm applied to the seven elementary indicators. Three clusters referring to (i) good, (ii) moderate, and (iii) low QoL were identified by minimizing the variance of the regions within each cluster and maximizing the variance of the regions outside each cluster, as reported below. A marked difference between low and good QoL clusters was found in the percentage of households reporting lack of telephone and absence of bath or shower, whereas the main difference between moderate and high QoL clusters was found in the percentage of households reporting pollution, crime, and other environmental issues caused by traffic or industry. On the contrary, the main difference between moderate and low QoL clusters was found in the percentage of households reporting vandalism or crime in the area. Regions above average QoL include Dytiki Makedonia, Sterea Ellada, Peloponnisos, Attiki, and Notio Aigaio. Ionia Nisia and Voreio Aigaio are in the middle of the rank. Finally, Thraki, Kentriki Makedonia, Thessalia, Ipeiros, Dytiki Ellada, Kriti, resulted to be in a position of below-average QoL.

Table 6. Individual indicators (after statistical transformation) and a composite index of subjective QoL based on the percentage of households reporting selected life conditions.

Region	Lack of Car or Can	Lack of Telephone	Lack of Home Computer	Lack of Bath or Shower	Lack of Hot Running Water	Pollution, Crime and Other Environmental Issues	Vandalism or Crime in the Area	QoL Composite Index
An. Makedonia	100	100	91.6	83.3	25	58.3	58.3	87.3
K. Makedonia	75	50	66.6	50.0	50.0	91.6	100.0	110.0
D. Makedonia	8.3	0.0	16.6	25.0	0.0	100.0	50.0	42.8
Thessalia	66.6	75.0	83.3	75.0	75.0	66.6	33.3	110.9
Ipeiros	83.3	58.3	75.0	91.6	91.6	0.0	83.3	120.1
Ionia Nisia	41.6	66.6	50.0	58.3	100	8.3	16.6	101.3
Dytiki Ellada	91.6	91.6	100	41.6	66.6	33.3	0.0	90.8
Sterea Ellada	0.0	25.0	25.0	8.3	8.3	16.6	41.6	26.9
Peloponissos	25.0	33.3	33.3	33.3	16.6	41.6	66.6	52.6
Attiki	16.6	16.6	0.0	0.0	41.6	83.3	91.6	77.6
Voreio Aigaio	50.0	41.6	41.6	66.6	58.3	25.0	25.0	75.8
Notio Aigaio	33.3	8.3	8.3	16.6	33.3	50.0	8.3	45.6
Kriti	58.3	83.3	58.3	100.0	83.3	75.0	75.0	129.6

Source: Authors' elaboration on official statistics released by Hellenic Statistical Authority, ELSTAT.

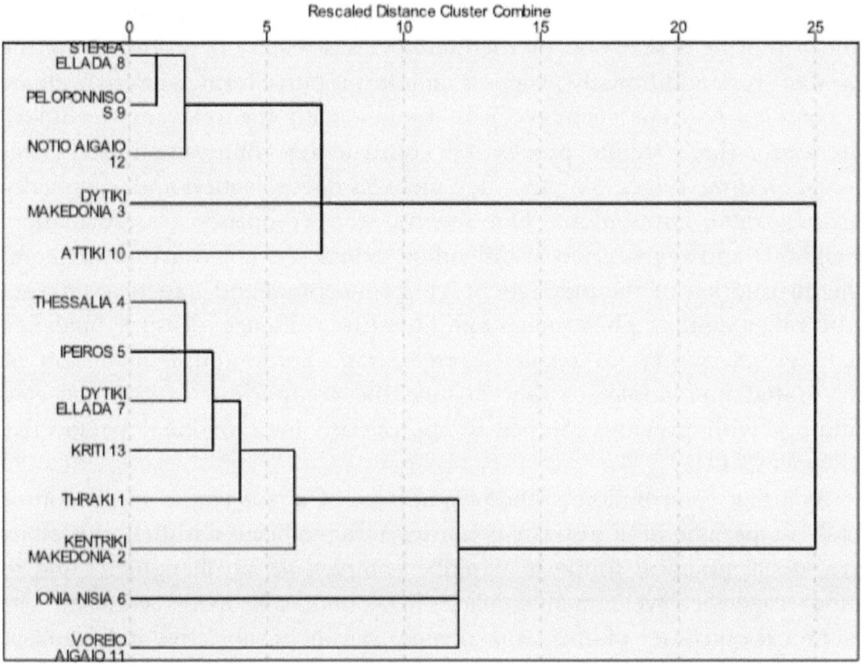

Figure 2. Similarity in the adopted (subjective) indicators of QoL across Greek regions using average linkage clustering.

Source: Authors' elaboration on official statistics released by Hellenic Statistical Authority, ELSTAT.

5. Discussion

A permanent assessment of socio-economic dynamics influencing QoL is imperative in contemporary societies (Faka et al. 2021). By using reliable information such as those derived from official statistics, composite indexes may provide, as a summary of elementary indicators, a more reliable knowledge of the current background context, feeding public debates and policy-making procedures (Salvati and Zitti 2009). A systemic exercise (*sensu* Chelleri et al. 2015) building up composite indexes of QoL is proposed in the present study, addressing complex QoL dimensions. This task entails the selection of target indicators, transformed appropriately to be compared using the same observation scale (Tarantola et al. 2002). The initial analysis' stage faces acceptability conditions and critical issues for the construction of composite indexes (Kosmas et al. 2016). An extensive review of advantages towards their meaningfulness in monitoring progress and performance of QoL across Greek regions has been provided (Di Feliciantonio et al. 2018).

By providing a comprehensive pathway to structure composite indicators and exploring a selection of methodologies at each operational step, the present work additionally proposes an interpretative framework to evaluate a complex concept such as QoL. In line with the relevant (technical) literature, the systemic process of constructing composite indexes was based on three stages. The first step includes data selection and, eventually, missing data imputation. The second step comprises transformation, weighting and aggregation of individual indicators. The third step explores the distribution of the index identifying conceptual and statistical linkages with other, similar, phenomena and literature evidence, possibly finalizing data visualisation and results' presentation. Future implementation of integrated methodologies that include the comparison of the collected findings with previous studies is appropriate in a regional perspective (Storper 2011).

As a first contribution to the pivotal issue of a 'subjective' evaluation of QoL in peripheral European countries using official statistics, this study provides a practical frame to describe complex social phenomena and to support specific developmental policies in economically divided contexts. The empirical outcomes of this work demonstrate how subjective indicators of life quality are essential for capturing the personal and emotional dimensions of well-being that objective measures often overlook (Diener 2009). While economic, health, and environmental statistics provide valuable insights, they do not fully reflect how individuals perceive and experience their own lives. Subjective indicators – such as life satisfaction, happiness, and perceived social support – offer a more holistic understanding of QoL by incorporating people's values, expectations, and lived experiences (Stiglitz, Sen, and Fitoussi 2009). Recognizing the importance of these indicators helps policymakers, researchers, and practitioners design more responsive and meaningful interventions that truly enhance human well-being.

6. Concluding Remarks

The present study demonstrates the feasibility of using official statistics in a subjective evaluation of QoL at a sufficiently disaggregated spatial scale, with the aim at estimating the importance of regional disparities and socio-demographic inequalities typical of a divided country like Greece (Monastiriotis 2009). Although typically grounded on a 'subjective' perspective, the academic contribution of this study to regional science lies in the fact that official statistics may provide reliable figures based on more precise sampling strategies and controlled procedures (standardized all over Europe) than primary surveys interviewing a restricted (and often less representative) sample of respondents. Official statistics should definitely perform – often in the framework of EU-SILC statistical vector – more refined indicators estimating subjective aspects of the 'quality of life' dimension, possibly allowing a direct comparison with other, close dimensions of well-being, such as personal incomes, employment rates, access to infrastructure, and health conditions (Tsakloglou 1990).

References

Anderberg, Michael R. 1973. "Hierarchical Clustering Methods." In *Cluster Analysis for Applications*, 131–155. Probability and Mathematical Statistics: A Series of Monographs and Textbooks, vol. 19. New York: Academic Press. https://doi.org/10.1016/B978-0-12-057650-0.50012-0.

Andrews, Frank M., and Stephen B. Withey. 1976. "Identifying and Mapping Concerns." In *Social Indicators of Well-Being: America's Perceptions of Life Quality*, 27–61. Boston, MA: Springer. https://doi.org/10.1007/978-1-4684-2253-5_2.

Bajocco, Sofia, Tomaso Ceccarelli, Daniela Smiraglia, Luca Salvati, and Carlo Ricotta. 2016. "Modeling the Ecological Niche of Long-Term Land Use Changes: The Role of Biophysical Factors." *Ecological Indicators* 60: 231–236. https://doi.org/10.1016/j.ecolind.2015.06.034.

Barro Robert J., and Xavier Sala-i-Martin. 2003. "*Economic Growth, 2nd Edition*," *MIT Press Books*, The MIT Press, 2nd edition.

Benyaklef, Mustafa. 1997. "Socio-economic Disparities in the Mediterranean." *Mediterranean Politics* 2 (1): 93–112. Doi:10.1080/13629399708414606.

Bericat, Eduardo, María Camarero, and María Luisa Jiménez-Rodrigo. 2019. "Towards a System of Indices on the Quality of European Societies (SIQES)." In *The Quality of European Societies*, edited by Eduardo Bericat and María Luisa Jiménez-Rodrigo, 1–28. *Social Indicators Research Series*, vol. 75. Cham: Springer. https://doi.org/10.1007/978-3-030-05023-8_1.

Booysen, Frederik. 2002. An Overview and Evaluation of Composite Indices of Development. *Social Indicators Research* 59, 115–151. https://doi.org/10.1023/A:1016275505152.

Campbell, Angus, Philip E. Converse, and Willard L. Rodgers. 1976. "Domains of Life Experience." In *Quality of American Life, The: Perceptions, Evaluations, and Satisfactions*, 61–96. Russell Sage Foundation. http://www.jstor.org/stable/10.7758/9781610441032.7.

Carlucci, Margherita, Efstathios Grigoriadis, Kostas Rontos, and Luca Salvati. 2017. "Revisiting a Hegemonic Concept: Long-Term 'Mediterranean Urbanization' in Between City Re-polarization and Metropolitan Decline." *Applied Spatial Analysis and Policy* 10 (3): 347–362. https://doi.org/10.1007/s12061-016-9186-2.

Cecchini, Massimo, Ilaria Zambon, Antonella Pontrandolfi, Rosario Turco, Andrea Colantoni, Anastasios Mavrakis, and Luca Salvati. 2019. "Urban Sprawl and the 'Olive' Landscape: Sustainable Land Management for 'Crisis' Cities." *GeoJournal* 84 (1): 237–255. https://www.jstor.org/stable/48699982.

Chelleri, Luca, Thorsten Schuetze, and Luca Salvati. 2015. "Integrating Resilience with Urban Sustainability in Neglected Neighborhoods: Challenges and Opportunities of Transitioning to Decentralized Water Management in Mexico City." *Habitat International* 48: 122–130. https://doi.org/10.1016/j.habitatint.2015.03.016.

Chorianopoulos, Ioannis, Georgios Tsilimigkas, Sotirios Koukoulas, and Thomas Balatsos. 2014. "The Shift to Competitiveness and a New Phase of Sprawl in the Mediterranean City: Enterprises Guiding Growth in Messoghia–Athens." *Cities* 39: 133–143. https://doi.org/10.1016/j.cities.2014.03.005.

Christopoulos, Dimitris K. 2004. "The Relationship between Output and Unemployment: Evidence from Greek Regions." *Papers in Regional Science* 83 (3): 611–620. https://doi.org/10.1111/j.1435-5597.2004.tb01928.x.

Ciommi, Mariateresa, Francesco M. Chelli, and Luca Salvati. 2019. "Integrating Parametric and Non-Parametric Multivariate Analysis of Urban Growth and Commuting Patterns in a European Metropolitan Area." *Quality & Quantity* 53: 957–979. https://doi.org/10.1007/s11135-018-0798-2.

Ciommi, Mariateresa, Francesco M. Chelli, Margherita Carlucci, and Luca Salvati. 2018. "Urban Growth and Demographic Dynamics in Southern Europe: Toward a New Statistical Approach to Regional Science" *Sustainability* 10(8): 2765. https://doi.org/10.3390/su10082765.

Costa, Daniel S. J. 2015. "Reflective, Causal, and Composite Indicators of Quality of Life: A Conceptual or an Empirical Distinction?" *Quality of Life Research* 24(9): 2057–2065. http://www.jstor.org/stable/44849302.

Cummins, Robert A. 1997. "Assessing Quality of Life." In *Quality of Life for People with Disabilities: Models, Research and Practice*, edited by Roy I. Brown, 2nd ed., 116–150. Cheltenham, UK: Stanley Thornes.

Di Feliciantonio, Cesare, and Luca Salvati. 2014. "Southern' Alternatives of Urban Diffusion: Investigating Settlement Characteristics and Socio-Economic Patterns in Three Mediterranean Regions." *Tijdschrift Voor Economische En Sociale Geografie* 106(4): 453–470. https://doi.org/10.1111/tesg.12102.

Di Feliciantonio, Cesare, Luca Salvati, Efthymia Sarantakou, and Kostas Rontos. 2018. "Class Diversification, Economic Growth and Urban Sprawl: Evidences from a Pre-Crisis European City." *Quality & Quantity* 52: 1501–1522. https://doi.org/10.1007/s11135-017-0532-5.

Diener, Ed. 2009. "Subjective Well-Being." In *The Science of Well-Being*, edited by Ed Diener, 11–58. *Social Indicators Research Series*, vol. 37. Dordrecht: Springer. https://doi.org/10.1007/978-90-481-2350-6_2.

Efklides, Anastasia, Maria Kalaitzidou, and George Chankin. 2003. "Subjective Quality of Life in Old Age in Greece: The Effect of Demographic Factors, Emotional State and Adaptation to Aging." *European Psychologist* 8 (3): 178–191. https://doi.org/10.1027/1016-9040.8.3.178.

European Commission: Eurostat. 2015. *Quality of Life – Facts and Views – 2015 Edition.* Luxembourg: Publications Office of the European Union. https://data.europa.eu/doi/10.2785/59737.

Faka, Antigoni, Kleomenis Kalogeropoulos, Thomas Maloutas, and Christos Chalkias. 2021. "Urban Quality of Life: Spatial Modeling and Indexing in Athens Metropolitan Area, Greece" *ISPRS International Journal of Geo-Information* 10 (5): 347. https://doi.org/10.3390/ijgi10050347

Freudenberg, Michael. 2003. "Composite Indicators of Country Performance: A Critical Assessment", *OECD Science, Technology and Industry Working Papers*, No. 2003/16, OECD Publishing, Paris, https://doi.org/10.1787/405566708255.

Funtowicz, Silvio O., and Jerome R. Ravetz. 1990. *Uncertainty and Quality in Science for Policy.* Dordrecht: Kluwer Academic Publishers. https://doi.org/10.1007/978-94-009-0621-1.

García-Bernabeu, Ana, José Manuel Cabello and Francisco Ruiz. 2022. "A Reference Point-Based Proposal to Build Regional Quality of Life Composite Indicators." *Social Indicators Research* 164: 11–30. https://doi.org/10.1007/s11205-021-02818-0.

Gavalas, Vasilis S., Kostas Rontos and Luca Salvati. 2014. "Who Becomes an Unwed Mother in Greece? Sociodemographic and Geographical Aspects of an Emerging Phenomenon." *Population Space and Place* 20: 250–263.

Greyling, Talita, and Fiona Tregenna. 2017. "Construction and Analysis of a Composite Quality of Life Index for a Region of South Africa." *Social Indicators Research* 131 (3): 887–930. https://www.jstor.org/stable/48715089.

Henning, Martin, Erik Stam, and Rik Wenting. 2013. "Path Dependence Research in Regional Economic Development: Cacophony or Knowledge Accumulation?" *Regional Studies* 47 (8): 1348–1362. Doi:10.1080/00343404.2012.750422.

Ivaldi, Enrico, Guido Bonatti and Riccardo Soliani. 2014. "Composite Index for Quality of Life in Italian Cities: An Application to URBES Indicators." *Review of Economics and Finance* 4: 18–32.

Jacobs, Rowena, Peter C. Smith, and Maria Goddard. 2004. *Measuring Performance: An Examination of Composite Performance Indicators.CHE Technical Paper Series 29*. York: Centre for Health Economics, University of York.

Kasimis, Charalambos. 2008. "Survival and Expansion: Migrants in Greek Rural Regions." *Population, Space and Place* 14 (6): 511–524. https://doi.org/10.1002/psp.513.

Keeney, Ralph L., and Howard Raiffa. 1993. "Aggregation of Individual Preferences." In *Decisions with Multiple Objectives: Preferences and Value Trade-Offs*, 515–548. Cambridge: Cambridge University Press. https://doi.org/10.1017/CBO9781139174084.012.

Kosmas, Costas, Mina Karamesouti, Kate Kounalaki, Vassilis Detsis, Penny Vassiliou, and Luca Salvati. 2016. "Land Degradation and Long-Term Changes in Agro-Pastoral Systems: An Empirical Analysis of Ecological Resilience in Asteroussia – Crete (Greece)." *CATENA* 147: 196–204. https://doi.org/10.1016/j.catena.2016.07.018.

Koutsampelas, Christos, and Panos Tsakloglou. 2013. The distribution of full income in Greece. *International Journal of Social Economics*. 40 (4): 311–330. https://doi.org/10.1108/03068291311305008.

Lamonica, Giuseppe R., Maria Cristina Recchioni, Francesco M. Chelli, and Luca Salvati. 2019. "The Efficiency of the Cross-Entropy Method When Estimating the Technical Coefficients of Input–Output Tables." *Spatial Economic Analysis* 15 (1): 62–91. Doi:10.1080/17421772.2019.1615634.

Maloutas, Thomas. 2007. "Segregation, Social Polarization and Immigration in Athens during the 1990s: Theoretical Expectations and Contextual Difference." *International Journal of Urban and Regional Research* 31 (4): 733–758. https://doi.org/10.1111/j.1468-2427.2007.00760.x.

Massart, D. Luc, and Leonard Kaufman. 1983. *The Interpretation of Analytical Chemical Data by the Use of Cluster Analysis*. New York: John Wiley & Sons.

Matsaganis, Manos, and Chrysoula Leventi. 2012. "The Distributional Impact of the Great Recession in Greece." Paper presented at the *32nd General Conference of the International Association for Research in Income and Wealth*, Boston, August 5–11.

Michalos, Alex C., Andrew Sharpe, and Nazeem Muhajarine. 2010. *An Approach to the Canadian Index of Wellbeing*. Toronto: Atkinson Charitable Foundation.

Monastiriotis, Vassilis. 2009. "Examining the Consistency of Spatial Association Patterns across Socio-Economic Indicators: An Application to the Greek Regions." *Empirical Economics* 37 (1): 25–49. https://doi.org/10.1007/s00181-008-0221-2.

———. 2011. Making geographical sense of the Greek austerity measures: compositional effects and long-run implications, *Cambridge Journal of Regions, Economy and Society*, 4 (3): 323–337. https://doi.org/10.1093/cjres/rsr026.

Nardo, Michele, Sebastien Tarantola, Andrea Saltelli, Chris Andropoulos, René Buescher, Giorgos Karageorgos, Anne Latvala, and Frédéric Noël. 2004. *The e-Business Readiness Composite Indicator for 2003: A Pilot Study*. Luxembourg: Office for Official Publications of the European Communities. JRC Scientific and Technical Reports. https://publications.jrc.ec.europa.eu/repository/handle/JRC28490.

Nardo, Michela, Michaela Saisana, Andrea Saltelli, and Stefano Tarantola. 2005. *Tools for Composite Indicators Building*. Ispra, Italy. Office for Official Publications of the European Communities. https://publications.jrc.ec.europa.eu/repository/handle/JRC31473.

Nunnally, Jum C. 1978. *Psychometric Theory*. 2nd ed. New York: McGraw-Hill.

OECD/European Union/EC-JRC. 2008. *Handbook on Constructing Composite Indicators: Methodology and User Guide*, OECD Publishing, Paris, https://doi.org/10.1787/9789264043466-en.

Papageorgiou, Kostas, Konstantinos Kassioumis, Vassilis Blioumis, and Athanasios Christodoulou. 2005. "Linking Quality of Life and Forest Values in Rural Areas: An Exploratory Study of Stakeholder Perspectives in the Rural Community of Konitsa, Greece." *Forestry: An International Journal of Forest Research* 78 (5): 485–499. https://doi.org/10.1093/forestry/cpi049.

Pappa, Evelina, Nick Kontodimopoulos, Angelos A. Papadopoulos, and Dimitris Niakas. 2009. "Assessing the Socio-Economic and Demographic Impact on Health-Related Quality of Life: Evidence from Greece." *International Journal of Public Health* 54 (4): 241–249. https://doi.org/10.1007/s00038-009-8057-x.

Pareto, Adriano. 2013. A Non-compensatory Composite Index for Measuring Well-being over Time. *Cogito Multidisciplinary research Journal*. 93–104.

Pili, Silvia, Efstathios Grigoriadis, Margherita Carlucci, Matteo Clemente, and Luca Salvati. 2017. "Towards Sustainable Growth? A Multi-Criteria Assessment of (Changing) Urban Forms." *Ecological Indicators* 76: 71–80. https://doi.org/10.1016/j.ecolind.2017.01.008.

Rodríguez-Pose, Andrés, and Ugo Fratesi. 2007. "Regional Business Cycles and the Emergence of Sheltered Economies in the Southern Periphery of Europe." *Growth and Change* 38 (4): 621–648. https://doi.org/10.1111/j.1468-2257.2007.00390.x.

Saisana, Michaela, Andrea Saltelli, and Stefano Tarantola. 2005. "Uncertainty and Sensitivity Analysis Techniques as Tools for the Quality Assessment of Composite Indicators." *Journal of the Royal Statistical Society. Series A (Statistics in Society)* 168, (2): 307–323. http://www.jstor.org/stable/3559964.

Saisana, Michaela, and Stefano Tarantola. 2002. "*State-of-the-Art Report on Current Methodologies and Practices for Composite Indicator Development*." *EUR 20408 EN*. Ispra, Italy: Joint Research Centre, European Commission.

Salvati, Luca, and Marco Zitti. 2009. "Substitutability and Weighting of Ecological and Economic Indicators: Exploring the Importance of Various Components of a Synthetic Index." *Ecological Economics* 68 (4): 1093–1099. https://doi.org/10.1016/j.ecolecon.2008.07.017.

Salvati, Luca, Marco Zitti, and Adele Sateriano. 2013. "Changes in City Vertical Profile as an Indicator of Sprawl: Evidence from a Mediterranean Urban Region." *Habitat International* 38: 119–125. https://doi.org/10.1016/j.habitatint.2012.05.006.

Salvati, Luca, and Margherita Carlucci. 2015. "Patterns of Sprawl: The Socioeconomic and Territorial Profile of Dispersed Urban Areas in Italy." *Regional Studies* 50 (8): 1346–1359. https://doi:10.1080/00343404.2015.1009435.

Salvati, Luca, Maria Teresa Ciommi, Pere Serra, and Francesco M. Chelli. 2019. "Exploring the Spatial Structure of Housing Prices under Economic Expansion and Stagnation: The Role of Socio-Demographic Factors in Metropolitan Rome, Italy." *Land Use Policy* 81: 143–152. https://doi.org/10.1016/j.landusepol.2018.10.030.

Stiglitz, Joseph E., Amartya Sen, and Jean-Paul Fitoussi. 2009. *Report by the Commission on the Measurement of Economic Performance and Social Progress*. Paris: Commission on the Measurement of Economic Performance and Social Progress.

Storper, Michael. 2011. "Why Do Regions Develop and Change? The Challenge for Geography and Economics." *Journal of Economic Geography* 11 (2): 333–346. http://www.jstor.org/stable/26162224.

Tarantola, Stefano, Michaela Saisana, Andrea Saltelli, Frank Schmiedel, and Nicholas Leapman. 2002. *Statistical Techniques and Participatory Approaches for the Compilation of the European Internal Market Index 1992–2001. EUR 20547 EN*. Luxembourg: European Commission, Joint Research Centre. https://publications.jrc.ec.europa.eu/repository/handle/JRC24431.

Tragaki, Alexandra, and Christos Bagavos. 2019. "Fertility Variations in the Recession Context: The Case of Greece." *Genus* 75 (1): 1–15. https://doi.org/10.1186/s41118-019-0066-x.

Tsakloglou, Panos. 1990. "Aspects of Poverty in Greece." *Review of Income and Wealth* 36 (4): 381–402. https://doi.org/10.1111/j.1475-4991.1990.tb00321.x

Zambon, Ilaria, Andrea Colantoni, Margherita Carlucci, Nathan Morrow, Adele Sateriano, and Luca Salvati. 2017. "Land Quality, Sustainable Development and Environmental Degradation in Agricultural Districts: A Computational Approach Based on Entropy Indexes." *Environmental Impact Assessment Review* 64: 37–46. https://doi.org/10.1016/j.eiar.2017.01.003.

Zuindeau, Bertrand. 2006. "Spatial Approach to Sustainable Development: Challenges of Equity and Efficacy." *Regional Studies* 40 (5): 459–470. https://doi.org/10.1080/00343400600757437.

Chapter 4

POPULATION DYNAMICS IN TURKEY (2000–2021): SPATIAL PATTERNS OF GROWTH, MIGRATION AND URBANIZATION

Bünyamin Özkür

1. Introduction

This chapter explores the population dynamics of Turkey between 2000 and 2021, a transformative period characterized by rapid urbanization, demographic transition and significant internal migration. Leveraging provincial-level data from the TURKSTAT, this analysis focuses on three interrelated dimensions of demographic change: total population growth, changes in urban-rural distribution and internal migration patterns. We examine these developments within the broader context of Turkey's spatial disparities and evolving socio-economic geography.

As of 2021, Turkey's population stood at approximately 84.7 million, having grown from 67.8 million in 2000 – a 25% increase (Yildiz et al. 2023; TURKSTAT 2000, 2021). This period witnessed a sharp demographic transition marked by declining fertility, increasing life expectancy and rapid population aging. Median age rose from the mid-20s to over 33.5 years by 2021, while the growth rate fell significantly (TURKSTAT Elderly Statistics, 2022). Regional disparities in demographic behaviour became more pronounced, with south-eastern provinces retaining high fertility and younger populations, while northern provinces exhibited aging and low fertility.

Urbanization trends were also dramatic: by 2021, over 80% of Turkey's population was classified as urban, compared to just 65% in 2000 (Akarca and Tansel 2015; TURKSTAT 2000, 2021). Migration, both voluntary and economically motivated, played a critical role in shaping this urban surge. Large metropolitan areas like Istanbul, Ankara and Izmir have become demographic magnets, increasingly populated by internal migrants (Kocaman 2008).

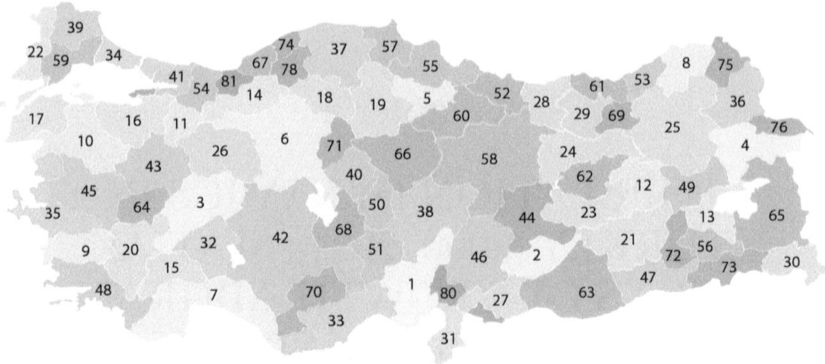

Figure 1. Administrative map of Turkey's 81 provinces (NUTS-3).

Source: Authors' elaboration using TURKSTAT administrative boundaries. Province codes correspond to Appendix A.

Figure 1 presents the administrative boundaries of Turkey's 81 provinces, which constitute the primary spatial units of analysis in this chapter. Understanding these provincial divisions is critical for interpreting the spatial disparities in demographic growth, urbanization and migration patterns discussed throughout. Each province is identified by a unique numerical code used in the subsequent maps and tables (see Appendix A).

2. Data and Methodology

This study employs a quantitative, longitudinal approach to analyse the spatial-temporal dynamics of the population in Turkey. The analysis is based on provincial-level (NUTS-3) data for the years 2000, 2011 and 2021, sourced exclusively from the TURKSTAT, the official state agency responsible for producing and disseminating national statistics.

The selection of these three specific years is methodologically deliberate. They represent critical junctures in Turkey's demographic data collection history, allowing for an examination of trends across distinct statistical paradigms:

- **2000**: This year marks the final traditional, nationwide General Population Census, providing a comprehensive baseline before major systemic changes.
- **2011**: This year offers a robust snapshot following the full implementation of the Address-Based Population Registration System (ADNKS) in 2007 and before the profound administrative changes of 2012.
- **2021**: This year provides the most recent comprehensive dataset available at the time of writing, allowing for an analysis of contemporary trends.

The primary indicators analysed include: Total Population (TP), Urban Population (UP), Rural Population (RP), Population Density (PD), calculated as persons per square kilometre), Urbanization Rate (U) and Net Internal Migration Rates. Provincial land area (km²) data, used in the calculation of population density, were obtained from the General Directorate of Mapping (Harita Genel Mudurlugu, 2014). These figures are derived from the 1:1,000,000-scale *Türkiye Mülki İdare Bölümleri Haritası* produced in 2014 and are indicative; they do not carry official legal status. Spatial analysis and visualization were conducted using the open-source software GeoDa to produce choropleth maps. A quantile classification method (dividing the 81 provinces into four equal groups) was used for all maps to facilitate consistent visual comparison of relative provincial rankings over time.

A critical appraisal of the data sources and their evolution is essential for a nuanced interpretation of the findings. The 21-year period under review is marked by two significant ruptures in data methodology and definition: the transition from a traditional census to a register-based system and the administrative redefinition of 'urban' and 'rural' spaces.

2.1. *The methodological rupture: from traditional census to the ADNKS*

The demographic data for the year 2000 are derived from a traditional, field-based General Population Census. This enumeration was conducted on a *de facto* basis, meaning individuals were counted at their physical location on a single, specific census day. This method, while standard for its era, was logistically intensive, expensive and provided only a periodic (decennial) snapshot of the population.

In 2006, with Law No. 5490, Turkey initiated a fundamental shift in its data collection infrastructure, culminating in the full implementation of the ADNKS in 2007. This system provides data for 2011 and 2021. ADNKS is a modern, register-based system that dynamically links two key administrative databases: the Central Civil Registration System (MERNIS), which contains personal identification information for all Turkish citizens and the National Address Database (NAD), which assigns a unique code to every residential address in the country.

This transition marked a shift from a *de facto* to a *de jure* population count. Under ADNKS, population is determined by an individual's "usual place of residence," defined as the address where they are officially registered and intend to live for more than six months. This system offers significant advantages in terms of cost-effectiveness and timeliness, providing annual population updates. However, this methodological shift creates a structural break in the time series between 2000 and the later years. Direct comparisons must be made with caution, as the change in definition can affect provincial counts. For example, a university student counted in their university city under the 2000 *de facto* census might be officially registered at their family's home under the *de jure* ADNKS, altering the population figures for both the sending and receiving provinces.

2.2. *The challenge of 'administrative urbanization': law no. 6360 and its statistical consequences*

The second and arguably more disruptive, rupture for longitudinal analysis is the Metropolitan Municipality Law No. 6360, enacted in 2012. This landmark legislation radically altered Turkey's administrative geography. It expanded the boundaries of the 16 existing metropolitan municipalities to encompass their entire respective provinces and created 14 new metropolitan municipalities, bringing the total to 30.

The most profound statistical consequence of this law was the abolition of the legal entity of over 16,000 villages and thousands of small towns (*belde*) within these 30 provinces. These settlements were administratively converted into 'neighbourhoods' (*mahalle*) of their respective district municipalities. This was a purely administrative reclassification, not a reflection of demographic change. As a result, millions of people living in functionally rural landscapes were, overnight, legally categorized as urban residents.

The phenomenon termed 'administrative urbanization' or 'paper urbanization' led to considerable distortion in official statistics. The national rural population share decreased from 22.7 per cent in 2012 to merely 8.7 per cent in 2013, a decline not due to real migration but rather a modification in classification.

This has two critical implications for this study. First, it renders the 'urbanization rate' for the 30 metropolitan provinces in 2021 meaningless for comparative purposes, as it is 100 per cent by legal definition. This explains the missing rural population data for these provinces in the 2021 dataset. The drastic visual shift towards high urbanization in the 2021 map (Figure 2) is therefore primarily a representation of this legal change, not organic urban growth. Second, this statistical reclassification risks making the needs of functionally rural populations within these large metropolitan provinces invisible to policymakers, a significant concern from a spatial justice perspective.

Recognizing this distortion, TURKSTAT has recently (in 2023) introduced a new three-tier settlement classification (densely populated, intermediate-density and thinly populated/rural) based on population density and contiguity,

(**a**)

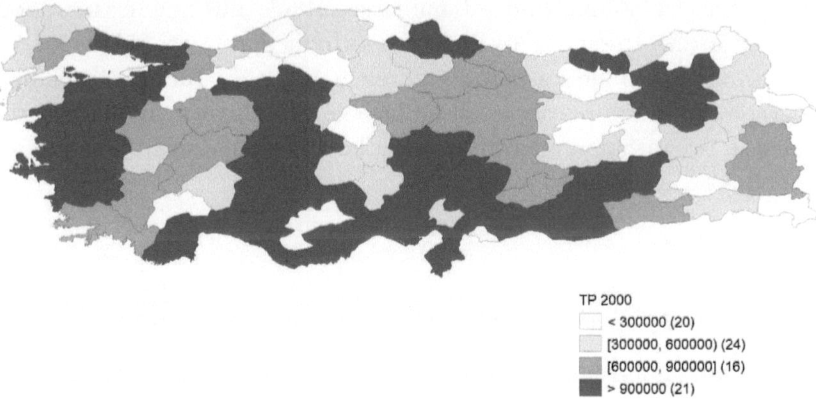

(**b**)

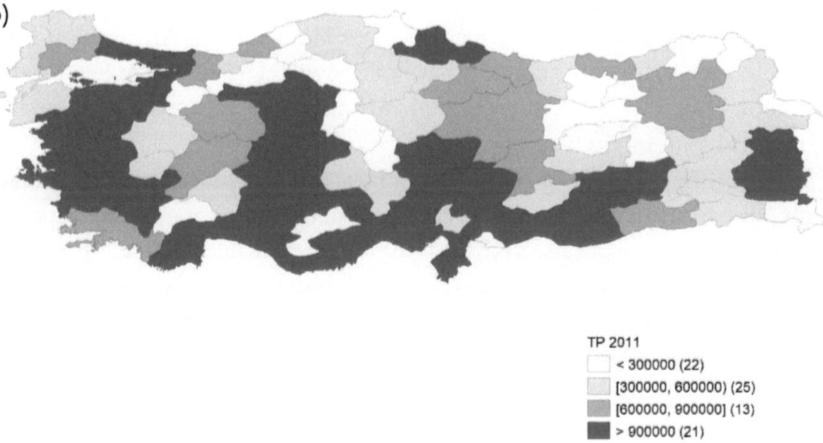

(**c**)

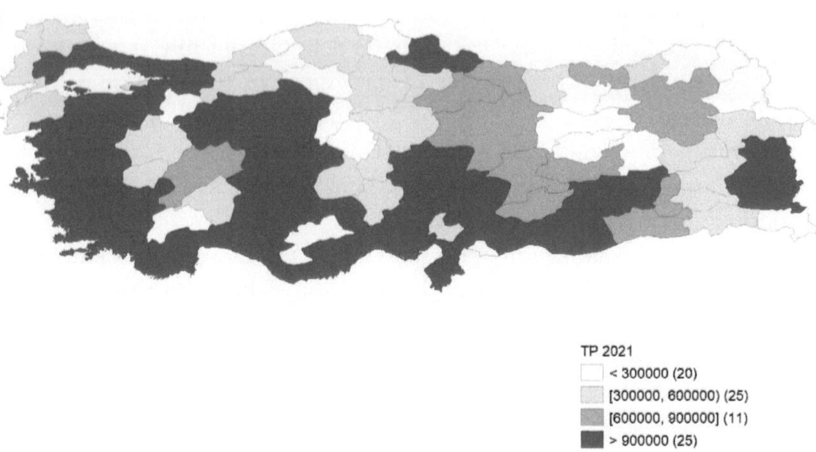

Figure 2. Total Provincial Population (2000 (a), 2011 (b), 2021 (c)) Choropleth maps classified into four quantiles show the distribution of provincial population across the three benchmark years, enabling consistent comparison of relative population size over time.

Source: TURKSTAT.

independent of administrative boundaries. While this new classification is not used in this study due to its recent introduction, it represents a crucial step towards a more accurate representation of Turkey's urban-rural geography in future research. All findings in this chapter, particularly those related to urbanization after 2012, must be interpreted with these profound methodological caveats in mind.

3. Findings

3.1. *National population growth and urbanization trends*

Turkey's total population grew by over 25 per cent between 2000 and 2021. Urban population share increased sharply, driven by both actual migration and administrative changes. Large provinces like Istanbul, Ankara and Izmir accounted for a disproportionate share of growth. Urbanization accelerated, especially after the 2012 Metropolitan Municipality Law, which reclassified many rural settlements as urban.

In 2000, high-population provinces (darkest tone) included Istanbul, Ankara, Izmir, Bursa and Adana, which already functioned as major metropolitan and economic centres (TURKSTAT 2000). Meanwhile, sparsely populated provinces like Ardahan, Bayburt and Tunceli remained in the lowest quantile, reflecting historical demographic stagnation.

By 2011, the spatial pattern of population concentration persisted, but with increased density in the Marmara and Mediterranean regions. Notably, provinces such as Antalya and Gaziantep rose in population class, indicating the effects of sustained migration and economic expansion (Kocaman 2008).

In 2021, Istanbul continues to dominate demographically, reaching almost 16 million residents. The number of provinces in the highest population quantile increased from 21 (2000) to 25 (2021), pointing to ongoing urban agglomeration in western and southern Turkey. Conversely, eastern provinces remained largely in the lowest class, evidencing entrenched spatial inequalities and limited demographic dynamism.

This tripartite visualization underscores a persistent west-east demographic divide and highlights how regional disparities in population size have proven resistant to change over the last two decades, despite national growth.

The maps depict the provincial distribution of total population in Turkey for the years 2000, 2011 and 2021, using quantile classification. Source data are derived from TURKSTAT's address-based population registration system.

Istanbul consistently holds the top rank across all three years – increasing from 10.0 million in 2000 to 15.8 million by 2021, alone comprising over 16 percent of the national population. It is followed by Ankara (5.7 million) and Izmir (4.4 million) in 2021, both maintaining their position within the highest population quantile.

In contrast, eastern provinces such as Bayburt (85,042), Tunceli (83,645) and Ardahan (94,932) remained in the lowest quantile throughout the entire period, reflecting demographic stagnation or net out-migration.

Rapid population growth is also evident in southern provinces like Antalya (rising from 1.5 million in 2000 to 2.6 million in 2021) and Sanliurfa (from 1.3 million to nearly 2.1 million), underscoring shifting internal dynamics. Despite these regional gains, the spatial concentration of population remains heavily skewed towards the Marmara and western coastal regions – reinforcing the long-standing west–east divide.

The spatial distribution demonstrates that while Turkey's overall population has grown from 67.8 million (2000) to 84.7 million (2021), this growth has been geographically uneven. High-population provinces have grown increasingly dominant, while many interior and eastern provinces remain demographically marginalized.

The persistence of this pattern raises important questions about regional disparities, infrastructure planning, and the long-term effects of urban primacy, especially as it relates to Istanbul's disproportionate role in national development.

3.2. *Urbanization patterns*

Between 2000 and 2021, urbanization spread to nearly all provinces. While western and coastal provinces were already highly urbanized, several central and eastern provinces became urbanized primarily due to administrative classification. This limits comparability across time but reflects the state's shifting spatial governance model.

Figure 3 displays choropleth maps of provincial urbanization rates in Turkey for the years 2000, 2011, and 2021, classified using quantile intervals. The top two maps correspond to 2000 and 2011, and the bottom map presents 2021. Data are derived from TURKSTAT census records and address-based registration systems (TURKSTAT 2000, 2011, 2021). In 2000, the spatial pattern of urbanization exhibited clear regional disparities: western provinces and large metropolitan areas (e.g. Istanbul, Ankara, Izmir) already had urbanization rates above 70 per cent, while many provinces in Eastern and South-eastern Anatolia remained predominantly rural. By 2011, this pattern had intensified slightly, with central provinces such as Konya and Kayseri exhibiting modest increases in urban rates, reflecting rural-to-urban migration trends and city-centred economic development.

(a)

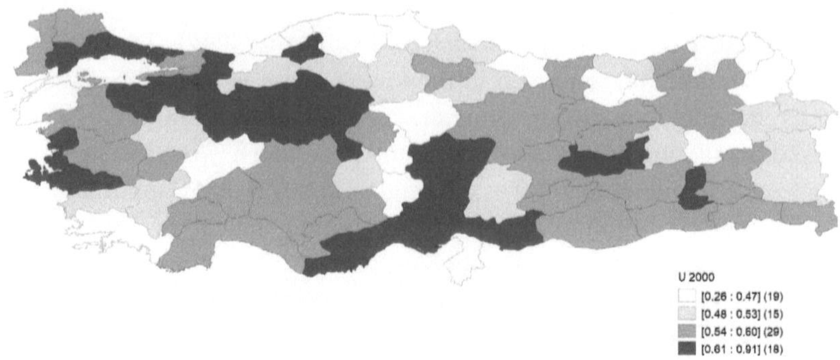

(b)

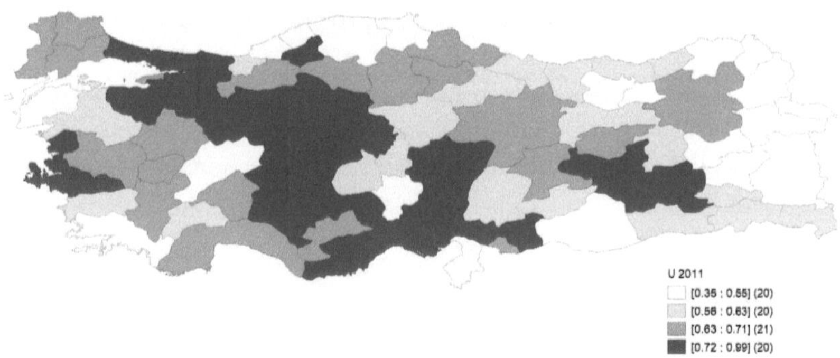

(c)

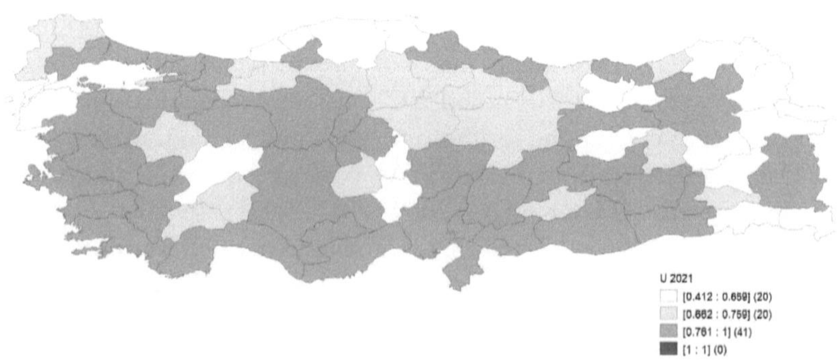

Figure 3. Urbanization Rate by Province (2000 (a), 2011 (b), 2021 (c)). Choropleth maps classified into quantile intervals illustrate the spatial distribution of provincial urbanization rates across three benchmark years.

Source: TURKSTAT.

The most dramatic transformation is visible in the 2021 map. Here, a substantial number of provinces show urbanization rates close to or at 100 percent, indicated by the dark orange shades dominating the map. This sudden jump is not entirely the result of organic urban growth but is largely attributable to the 2012 Metropolitan Municipality Law (No. 6360), which reclassified entire provinces as metropolitan areas, thereby redefining rural settlements as urban neighbourhoods (Savas-Yavuzcehre 2016). For instance, Sanliurfa and Van – previously classified as mixed or rural-heavy – appear fully urbanized by 2021. This 'administrative urbanization' reclassified rural settlements as urban within metropolitan boundaries, altering the statistical urbanization rate without corresponding demographic change. This transformation raises critical questions for interpreting longitudinal urbanization trends. While the maps indicate a spatial convergence towards high urbanization, the underlying drivers include both demographic changes and political-administrative reforms. Caution is therefore warranted when comparing pre- and post-2012 urbanization figures. Still, the trend reflects Turkey's long-term urban trajectory: consolidation of population in urban centres, marginalization of rural peripheries and the increasing dominance of metropolitan governance frameworks.

3.3. *Population density shifts*

Population density rose in nearly all regions, with striking increases in Marmara, Aegean and Mediterranean coastal provinces. In contrast, many eastern provinces remain sparsely populated. Rising densities in western Turkey put strain on urban infrastructure and regional planning.

Figure 4 displays the provincial distribution of population density (persons per km^2) in Turkey for the years 2000, 2011 and 2021, using quantile classification across four intervals. The visualization highlights long-term spatial continuity in the geography of density, with the western and coastal provinces consistently showing higher concentrations of population compared to the interior and eastern regions.

In 2000, the densest provinces (>100 persons/km^2) were concentrated around Istanbul, Kocaeli, Izmir, Bursa and Gaziantep, reflecting early agglomeration patterns around industrial and urban cores (TURKSTAT 2000). The lowest-density provinces (<40 persons/km^2) were largely located in the Eastern Anatolia Region, such as Tunceli, Ardahan and Bayburt.

By 2011, the maps show only moderate shifts in the spatial pattern. The number of provinces in the highest density class increased, notably in the southern belt (e.g. Hatay and Adana) and central Anatolian provinces that experienced regional growth due to infrastructure investments and internal

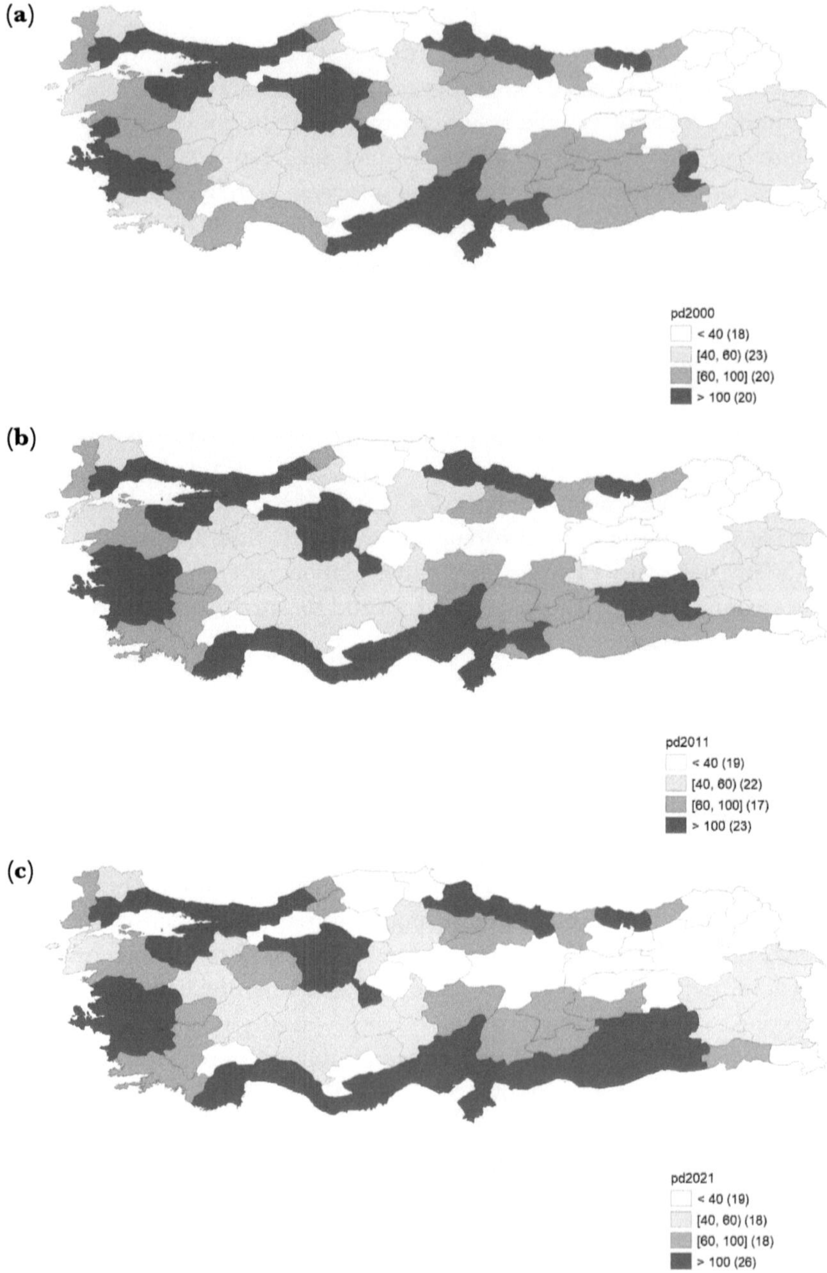

Figure 4. Population Density by Province (2000 (a), 2011 (b), 2021 (c)) Choropleth maps classified into four quantiles show the spatial distribution of provincial population density (persons per km²) for the three benchmark years, enabling consistent comparison of relative density levels over time.

Source: TURKSTAT.

migration (Karaalp-Orhan 2020). The low-density east persisted, suggesting minimal in-migration or local growth.

In 2021, the high-density provinces expanded further to include parts of Southeast Anatolia (e.g. Sanliurfa) and western growth centres such as Tekirdag and Aydin remained in the highest bracket (>100 persons/km²). Istanbul's density alone exceeded 3,000 persons/km² – far outpacing other provinces and emphasizing the growing demographic centrality of the Marmara Region (Yildiz et al. 2023).

The constancy of the east–west divide during the 21-year period is particularly noteworthy. Despite substantial population growth in Turkey, spatial inequality in density persisted, highlighting persistent structural imbalances and the constraints of regional convergence.

3.4. *Migration patterns*

Migration trends show strong east-to-west movements, especially among working-age populations. Provinces like Istanbul, Antalya and Kocaeli were top gainers, while eastern provinces such as Yozgat and Ardahan experienced sustained out-migration. These flows have major implications for service provision, labour markets and local economies.

This three-panel map (Figure 5) displays the spatial distribution of net internal migration rates (per 1,000 population) across Turkey's 81 provinces for three benchmark periods: 1995–2000, 2010–2011 and 2020–2021. The maps classify provinces into four categories based on net migration rate per 1,000 population, allowing for comparative analysis over time.

In 1995–2000, migration patterns reveal a clear east-to-west flow. Eastern provinces such as Ağrı, Muş, Bitlis and Şırnak experienced high net out-migration rates (below –40 per 1,000), indicating strong push factors such as economic stagnation, insecurity and limited infrastructure (Erder and Kaska 2003; Icduygu et al. 2001). Conversely, western urban-industrial provinces like Istanbul, Kocaeli, Aydın and Antalya stood out as strong in-migration hubs, with net migration rates exceeding +10 per 1,000.

By 2010–2011, the picture changes moderately. The number of provinces with extreme net losses (<–40 per 1,000) significantly declined, suggesting a broad stabilization of migration flows. The dominant pattern becomes one of mild migration, as most provinces clustered around the –10 to +10 per 1,000 category. This reflects increasing urban saturation in western metropolises and improving conditions in smaller Anatolian cities, partially driven by state-led regional development programs (DPT 2006).

In 2020–2021, despite overall population growth, the net migration map shows a more diffused pattern. Out-migration still affects parts of the

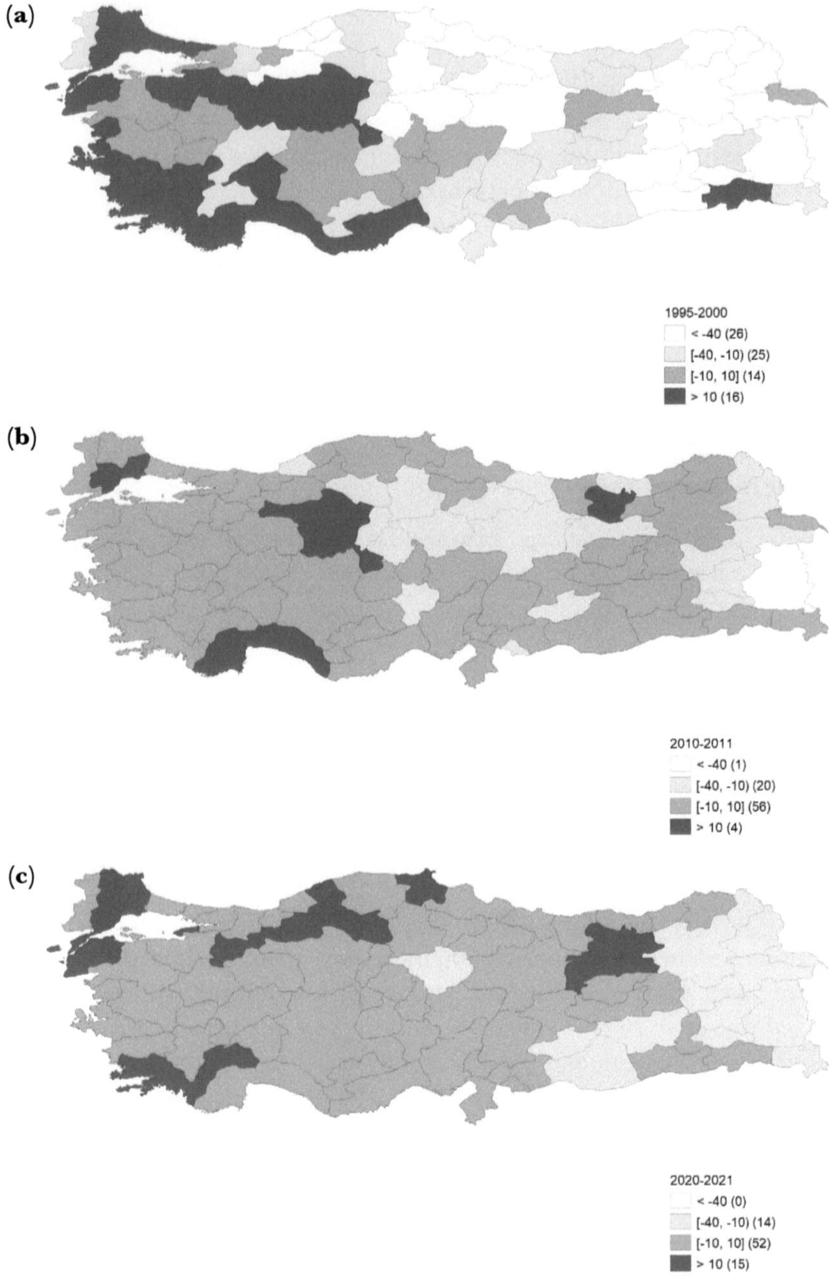

Figure 5. Net Internal Migration Rate per Province in Turkey (1995–2000 (a), 2010–2011 (b), 2020–2021 (c)). Provinces are classified into four categories by net migration rate (per 1,000 population), enabling temporal comparison across three benchmark periods.

Source: TURKSTAT.

eastern Black Sea and north-eastern Anatolia (e.g. Bayburt, Gumushane), while some smaller central provinces (e.g. Corum, Tokat) also show net losses. However, migration hotspots persist in the west and south (e.g. Antalya, Tekirdag), confirming the continued pull of economically vibrant regions (Karaalp-Orhan 2020). Interestingly, migration gains are also recorded in emerging secondary cities, such as Ordu and Balıkesir, suggesting growing polycentricity.

Overall, the maps confirm long-term trends of internal migration as a mechanism of regional adjustment in Turkey. Although extreme migration flows from the east have declined, inter-provincial disparities remain and internal mobility continues to redistribute population towards coastal and industrial zones. These findings support previous research that identifies internal migration as both a response to and a driver of spatial inequality in Turkey (Gursoy and Badur 2022; Yildiz et al. 2023).

The histograms in Figure 6 illustrate the distribution of net migration rates (per 1,000 population) across provinces for three benchmark periods. In 1995–2000, the distribution is sharply right-skewed, with many provinces

(a)

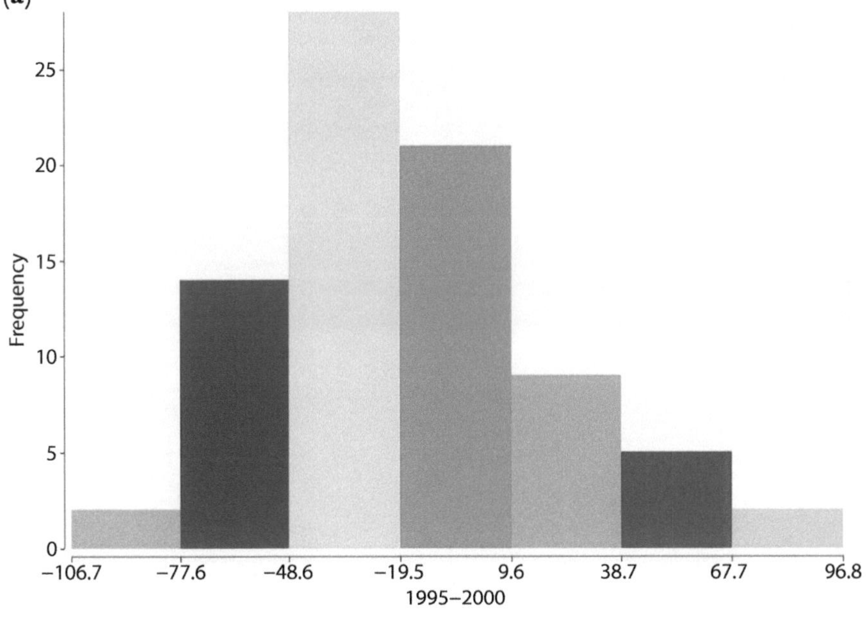

Figure 6. Histograms of provincial net migration rates (1995–2000 (a), 2010–2011 (b), 2020–2021 (c)). Distributions of net migration rates (per 1,000 population) across provinces at three time points reveal a pronounced right-skew in 1995–2000 and greater centralization by 2020–2021.

Source: TURKSTAT.

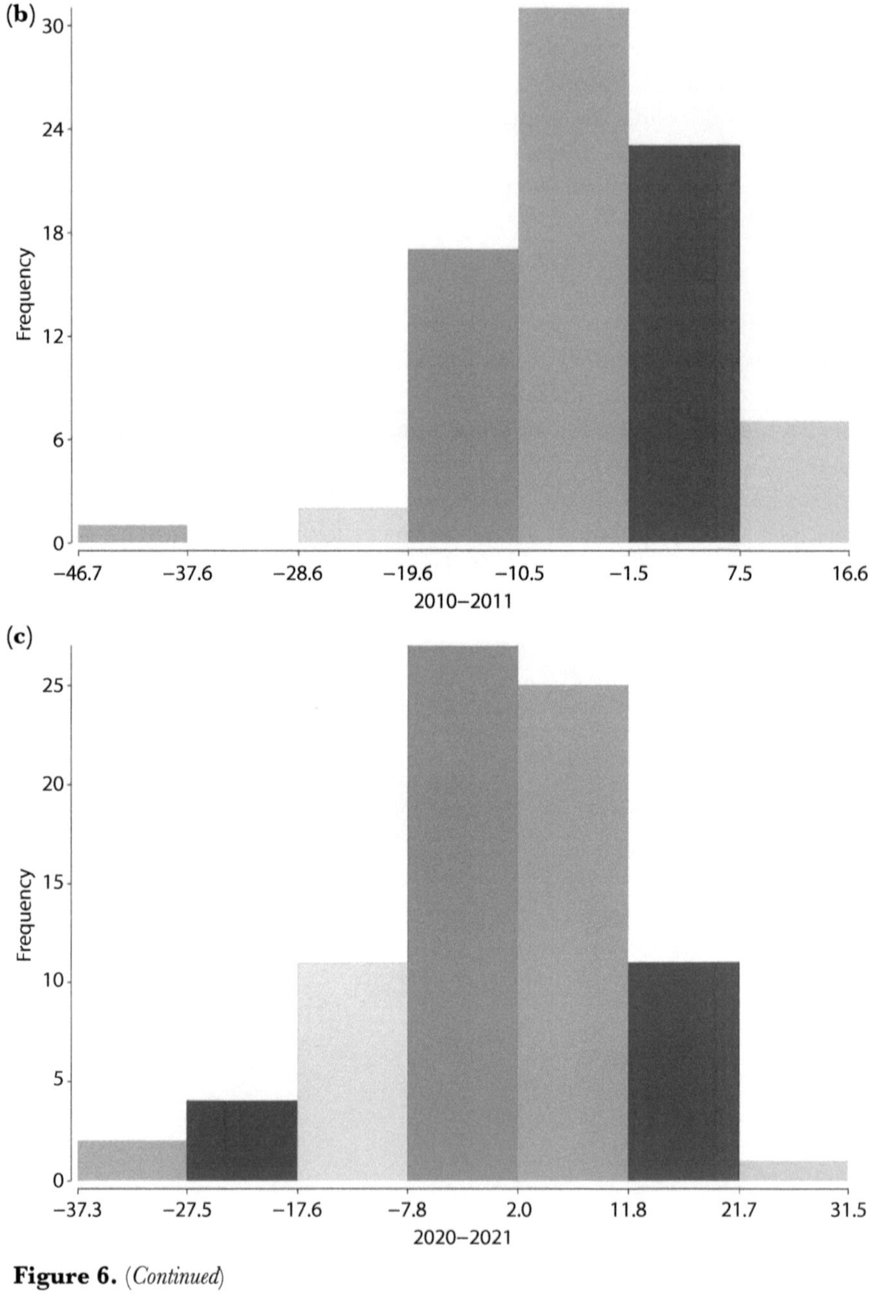

Figure 6. (*Continued*)

experiencing substantial out-migration and only a few attracting significant in-migration. By 2020–2021, the distribution becomes more centralized around zero, indicating a modest convergence in migration dynamics across provinces. Most provinces now cluster near migration balance, while a handful (e.g. Istanbul, Kocaeli, Tekirdag) continue to register high in-migration and others (e.g. Yozgat, Kars, Agri) persistently record strong out-migration. These patterns point to ongoing – though moderated – regional disparities in internal mobility (TURKSTAT 2023).

4. Discussion

The demographic patterns observed between 2000 and 2021 underscore a central and enduring feature of Turkey's geography: a profound and persistent west–east divide. This is not merely an economic lag but a structural chasm that decades of regional development policy have failed to close. Since the 1960s, state-led initiatives, most notably the Southeastern Anatolia Project (GAP), have aimed to reduce interregional disparities through large-scale infrastructure investments (Kurt 2013). However, the continued out-migration from eastern provinces suggests these top-down projects have been insufficient to generate the sustainable, diversified economic growth needed to alter migration incentives fundamentally (Eraydin 2015). The result is a self-perpetuating cycle of cumulative causation: western metropolitan centres like Istanbul, Kocaeli and Bursa benefit from agglomeration economies, attracting investment and high-skilled labour, which in turn makes them even more attractive (World Bank 2015). This process siphons human and financial capital from the periphery, eroding the tax base and growth potential of eastern regions and creating a deep-seated issue of spatial justice (Karaalp-Orhan 2020).

However, new dynamics have complicated the traditional east-to-west migration narrative. The arrival of over 3.6 million Syrian refugees since 2011 represents a major demographic shock (UNHCR n.d.). While many refugees eventually moved to western cities, their initial concentration in south-eastern provinces created a significant labour supply shock. This has been shown to deter the internal migration of native Turks into these areas, as competition for low-skilled jobs and resources like housing increased (Akgunduz et al. 2015). This factor adds a crucial layer of complexity, disrupting the simple core–periphery model by showing how international migration can reshape a country's internal mobility patterns.

Simultaneously, Turkey's urban system is evolving beyond a simple monocentric model focused on a few megacities. The period has seen the rise of dynamic secondary cities, the 'Anatolian Tigers', such as Gaziantep,

Kayseri and Konya, which have become powerful regional economic engines and migration hubs in their own right (World Bank 2015). These cities attract significant migration from their surrounding hinterlands, suggesting a shift towards a more polycentric urban system where development is more distributed (Sat 2018). Furthermore, evidence of counter-urbanization – flows from large metropolitan areas to smaller cities and rural settlements, driven by QoL considerations and accelerated by the COVID-19 pandemic – indicates that mobility is becoming more multidirectional.

A final complicating factor is the statistical artefact created by the 2012 Metropolitan Municipality Law (No. 6360). This reform abolished the legal status of villages and many small towns ('belde') in 30 metropolitan provinces and converted them into urban 'neighbourhoods', producing an immediate statistical drop in the rural population (Kazan 2017; Kizilaslan et al. 2016). This 'administrative urbanization' means that the dramatic increase in urbanization rates in the 2021 data are primarily a reflection of legal reclassification, not organic growth. It also obscures functionally rural populations inside metropolitan boundaries, complicating needs assessment and policy design; Istanbul's 'rural neighbourhoods' remain socio-spatially rural despite their urban label (Yaman Kocadagli 2022).

In summary, Turkey's demographic geography is the outcome of a complex interplay of factors: long-standing structural inequalities, the limited success of centralized development policies, a major international migration shock and the emergence of a more complex, polycentric network of cities. Internal migration continues to be a rational response to spatial inequality, but the pathways of that migration are becoming more diverse and nuanced than ever before.

5. Concluding Remarks

The era from 2000 to 2021 was characterized by geographically imbalanced growth, exacerbating Turkey's spatial disparities. The national population increased by 25%, rising from 67.8 million to 84.7 million, with this growth predominantly occurring in select urban areas in the west and south (Yildiz et al. 2023). Provinces such as Istanbul, Ankara and Izmir flourished, with Istanbul's population density escalating from roughly 1,835 to exceeding 3,000 persons per km^2 (TURKSTAT 2000, 2021). In sharp contrast, peripheral provinces in the east and interior, like Yozgat and Ardahan, underwent persistent depopulation, with Yozgat's population declining by about 40 per cent (TURKSTAT 2000, 2021).This has established a dichotomy of two Turkeys: one urban, expanding and globally integrated and the other rural, declining and at risk of lagging behind.

These trends face a new and challenging demographic backdrop in the future. The era of rapid population growth is ending, as Turkey confronts the dual pressures of a rapidly aging population and a fertility rate that has fallen to 1.51, well below replacement level (TURKSTAT 2023). Projections indicate that the median age will continue to rise and the country's "demographic window of opportunity" – the period with a favourable ratio of working-age to dependent population – is expected to close in the early 2030s (TURKSTAT 2023). This impending shift will place significant strain on social security and healthcare systems and could act as a headwind to future economic growth.

The consequences of this unbalanced population distribution extend to the environment. In the West, hyper-urbanization has exacted a heavy toll. The relentless expansion of Istanbul has led to the fragmentation and destruction of the vital Northern Forests, which are critical for the city's air and water quality (Saribas and Balcik 2022; Karaburun et al. 2010). This urban sprawl contributes to increased air pollution and exacerbates the urban heat island effect. Conversely, the depopulation of rural regions has its own environmental costs. The abandonment of agricultural lands leads to the loss of traditional, biodiverse landscapes and cultural heritage, while neglected fields and forests increase the risk of uncontrolled wildfires (MAPFRE 2021; Diker 2022).

To tackle these complicated issues related to people, the economy and the environment, we need a united and future-focused policy plan based on fairness in space and sustainability, following the UN Sustainable Development Goals 10 (Reduced Inequalities) and 11 (Sustainable Cities and Communities) (UN 2015; OECD 2022). A cohesive national strategy should prioritize:

1. Fostering a Polycentric Urban Network: Shift policy focus from managing megacities to actively promoting secondary cities as alternative centres of economic gravity. Strategic investments in the 'Anatolian Tigers' can create a more balanced and resilient national urban system (World Bank 2015; Sat 2018).
2. Pursuing Integrated Regional Development: Move beyond top-down infrastructure projects to address the root causes of out-migration. This requires integrated strategies that build local human capital, improve governance and create a favourable environment for private investment in lagging regions (Kurt 2013).
3. Implementing Sustainable Urban Planning: In high-growth metropolitan areas, enforce greenbelt policies to protect ecosystems, invest in public transportation to combat pollution and promote affordable housing to ensure social inclusion (UN 2015).

4. Recognizing and Supporting Rurality: Develop policies that support rural livelihoods and protect agricultural land and cultural heritage, acknowledging that functionally rural areas persist even within metropolitan provinces (Kosker-Yaman and Kutlar 2017).
5. Preparing for an Aging Society: Proactively plan for the impacts of an aging population by reforming pension and healthcare systems and developing strategies to maintain economic dynamism with a changing workforce (TURKSTAT 2024; OECD 2016).

By adopting such a spatially just and sustainable approach, Turkey can mitigate the pressures of unbalanced development and work towards a future where geography no longer determines destiny, ensuring that its population dynamics contribute to, rather than hinder, long-term national prosperity.

Appendix A. List of Turkish Provinces by Numerical Code

Code	Province	Code	Province	Code	Province
1	Adana	2	Adiyaman	3	Afyon
4	Agri	5	Amasya	6	Ankara
7	Antalya	8	Artvin	9	Aydin
10	Balikesir	11	Bilecik	12	Bingöl
13	Bitlis	14	Bolu	15	Burdur
16	Bursa	17	Canakkale	18	Cankiri
19	Corum	20	Denizli	21	Diyarbakir
22	Edirne	23	Elazig	24	Erzincan
25	Erzurum	26	Eskisehir	27	Gaziantep
28	Giresun	29	Gümüshane	30	Hakkari
31	Hatay	32	Isparta	33	Mersin
34	Istanbul	35	Izmir	36	Kars
37	Kastamonu	38	Kayseri	39	Kirklareli
40	Kirsehir	41	Kocaeli	42	Konya
43	Kütahya	44	Malatya	45	Manisa
46	Kahramanmaras	47	Mardin	48	Mugla
49	Mus	50	Nevsehir	51	Nigde
52	Ordu	53	Rize	54	Sakarya
55	Samsun	56	Siirt	57	Sinop
58	Sivas	59	Tekirdag	60	Tokat
61	Trabzon	62	Tunceli	63	Sanliurfa

Code	Province	Code	Province	Code	Province
64	Usak	65	Van	66	Yozgat
67	Zonguldak	68	Aksaray	69	Bayburt
70	Karaman	71	Kirikkale	72	Batman
73	Sirnak	74	Bartin	75	Ardahan
76	Igdir	77	Yalova	78	Karabuk
79	Kilis	80	Osmaniye	81	Duzce

References

Akarca, T. Ali, and Aysit Tansel. 2015. "Impact of Internal Migration on Political Participation in Turkey." *IZA Journal of Migration* 4(1): 1–14. https://doi.org/10.1186/s40176-014-0025-4.

Akgündüz, Yusuf Emre, Marcel van den Berg, and Wolter Hassink. 2015. *The impact of refugee crises on host labor markets: The case of the Syrian refugee crisis in Turkey.* IZA Discussion Paper No. 8841, Institute of Labor Economics (IZA), Bonn. https://docs.iza.org/dp8841.pdf.

Diker, İrem. 2022. "Principles for the Conservation of Depopulated Rural Heritage Sites: The Case of Dereköy on Gökçeada (Imbros)." MS Thesis. Middle East Technical University (Turkey). ProQuest (3122679256).

DPT (Türkiye Devlet Planlama Teşkilatı). 2006. Dokuzuncu Kalkınma Planı 2007–2013. Ankara: DPT. https://www.sbb.gov.tr/wp-content/uploads/2022/07/Dokuzuncu_Kalkinma_Plani-2007-2013.pdf.

Eraydin, Ayda. 2015. "Attributes and Characteristics of Regional Resilience: Defining and Measuring the Resilience of Turkish Regions." *Regional Studies* 50(4): 600–614. https://doi.org/10.1080/00343404.2015.1034672.

Erder, Sema, and Selmin Kaska. 2003. *Irregular Migration and Trafficking in Women: The Case of Turkey.* The International Organization for Migration (IOM) Publications. http://www.iom.int/DOCUMENTS/PUBLICATION/EN/Irregular%5Fmig%5Fin%5Fturkey.pdf.

Gürsoy, Furkan, and Bertan Badur. 2022. "Investigating internal migration with network analysis and latent space representations: an application to Turkey." *Social Network Analysis and Mining* 12(1): 150. doi:10.1007/s13278-022-00974-w.

Harita Genel Müdürlüğü. 2014. *İl ve ilçe yüzölçümleri* [Provincial and District Surface Areas]. Accessed on July 10, 2025. https://www.harita.gov.tr/urun/il-ve-ilce-yuzolcumleri/176.

İçduygu, Ahmet, İbrahim Sirkeci, and Gülnur Muradoğlu. 2001. Socio-economic Development and International Migration: A Turkish Study. *International Migration* 39(4): 39–61. https://doi.org/10.1111/1468-2435.00162.

Karaalp-Orhan, S. Hacer. 2020. Regional Disparities in Turkey: A Socio-Economic Perspective. *European Journal of Sustainable Development* 9(3): 103–115. https://doi.org/10.14207/ejsd.2020.v9n3p103.

Karaburun, Ahmet, Ali Demirci, and Suen I-shian. 2010. "Impacts of urban growth on forest cover in Istanbul (1987–2007)." *Environmental Monitoring and Assessment* 166(1–4): 267–277. doi:10.1007/s10661-009-1000-z.

Kazan, Ahmet. 2017. The Effects of the Law no 6360 on the Financial Structures of Metropolitan Municipalities. *Sayıştay Dergisi* 104: 87–106. https://www.sayistay.gov.tr/files/4465_der104m5.pdf.

Kızılaslan, Halil, Tayfur Ünal, and Nuray Kızılaslan. 2016. Effects of new metropolitan law No. 6360 to rural development in Turkey. *Journal of New Theory* (13), 76–85.

Kocaman, Tuncer. 2008. *Türkiye'de iç göçler ve göç edenlerin nitelikleri (1965–2000)*. Ankara: Devlet Planlama Teşkilatı.

Köşker-Yaman, Ayşe, and İlkay Kutlar. 2017. "Potential Impacts of the Metropolitan Municipality Law No. 6360 on Rural Life: The Case of Kocaeli". *Turkish Journal of Agriculture – Food Science and Technology* 5(11):1295–1300. https://doi.org/10.24925/turjaf.v5i11.1295-1300.1347.

Kurt, Çiğdem. 2013. "The Impact of the Southeast Anatolia Project (Gap) on Displaced Families: Household Livelihoods and Gender Relations." PhD diss., University of Newcastle Upon Tyne (United Kingdom). ProQuest (1797588701).

MAPFRE. 2021. "Why Should Rural Depopulation Be Avoided?" MAPFRE Insights. Accessed on January 19, 2025. https://www.mapfre.com/en/insights/sustainability/rural-depopulation/.

Sarıbaş, Büşra, and Filiz Bektaş Balçık. 2022. "Determination of the Impact of Urbanization in Istanbul Northern Forests by Remote Sensing" *Environmental Sciences Proceedings* 22(1): 57. https://doi.org/10.3390/IECF2022-13059.

Sat, N. Aydan. 2018. Polycentricity in a developing world: A micro-regional analysis for morphological polycentricity in Turkey. *GeoScape* 12(2): 64–75. https://doi.org/10.2478/geosc-2018-0007.

Savaş-Yavuzçehre, Pınar. 2016. "The Effects of the Law No. 6360 on Metropolitan Municipality System in Turkey." *European Scientific Journal*, Special Edition (August). https://eujournal.org/index.php/esj/issue/view/243.

The Organization for Economic Cooperation and Development (OECD). 2016. *OECD Economic Surveys: Turkey 2016*, OECD Publishing, Paris, https://doi.org/10.1787/eco_surveys-tur-2016-en.

The Organization for Economic Cooperation and Development (OECD). 2022. "The short and winding road to 2030: Measuring distance to the SDG targets", *OECD Policy Insights on Well-being, Inclusion and Equal Opportunity*. Vol. 5, OECD Publishing, Paris, https://doi.org/10.1787/3326f232-en.

TURKSTAT (Turkish Statistical Institute). (2000). *Population of provinces by census years*. https://data.tuik.gov.tr/Kategori/GetKategori?p=Population-and-Demography-109.

TURKSTAT (Turkish Statistical Institute). (2011). *Address-based population registration system (ABPRS)*. https://data.tuik.gov.tr/Kategori/GetKategori?p=Population-and-Demography-109.

TURKSTAT (Turkish Statistical Institute). (2021). *Address-based population registration system (ABPRS)*. https://data.tuik.gov.tr/Kategori/GetKategori?p=Population-and-Demography-109.

TURKSTAT (Turkish Statistical Institute). *Elderly Statistics*, 2022. Ankara: TURKSTAT, 2022. https://data.tuik.gov.tr/Bulten/Index?dil=2&p=Elderly-Statistics-2022-49667.

TURKSTAT (Turkish Statistical Institute). 2023. *Population projections*, 2023–2100. https://data.tuik.gov.tr/Bulten/Index?p=Population-Projections-2023-2100-53699&dil=2.

TURKSTAT (Turkish Statistical Institute). 2024. *Elderly statistics, 2024*. https://data.tuik.gov.tr/Bulten/Index?p=Elderly-Statistics-2024-54079.

UN (United Nations). 2015. *Transforming Our World: The 2030 Agenda for Sustainable Development*. New York, United Nations. https://sdgs.un.org/2030agenda.

UNHCR (United Nations High Commissioner for Refugees). n.d. *Operational Data Portal – Syria Situation*. UNHCR. Accessed August 1, 2025. https://data.unhcr.org/en/situations/syria.

World Bank. 2015. *Rise of the Anatolian Tigers: Turkey Urbanization Review, Main Report. Policy Brief.* Washington, DC: World Bank. https://hdl.handle.net/10986/22388.

Yaman Kocadağlı, Aylin. 2022. İstanbul'un Kırsal Mahalleleri: Güncel Yasal Düzenlemeler Kapsamında Bir Durum Analizi. *Eurasian Academy of Sciences Social Sciences Journal* 40: 144–167. https://socialsciences.eurasianacademy.org/dergi/istanbulun-kirsal-mahalleleri-guncel-yasal-duzenlemeler-kapsaminda-bir-durum-analizi202203.pdf.

Yıldız, Dilek, Tuğba Adalı, and Caner Özdemir. 2023. "An Overview of Türkiye's Population on the 100th Anniversary of the Republic." *Population & Societies* 608(2): 1–4. https://doi.org/10.3917/popsoc.608.0001.

Chapter 5

SOCIAL INEQUALITIES, RURAL POVERTY AND LANDSCAPE DEGRADATION IN ADVANCED ECONOMIES: A REGIONAL PERSPECTIVE

Bünyamin Özkür, Rosanna Salvia, Giovanni Quaranta, Marco Maialetti, Ioannis Konaxis, Bogdana Nosova, Jesus Rodrigo Comino and Luca Salvati

1. Introduction

Environmental changes resulting from biophysical factors and human activities have generated significant interest in the social sciences, particularly regarding the potential impacts of landscape degradation on population dynamics and economic systems, as well as societal responses to these changes (e.g. Hussen 2004). Uncertainty and risk are central to the processes of social dynamics and political actions (Nocenzi 2005). Meanwhile, the economic discourse on sustainable growth remains prevalent, marked by the intricate interplay of ecological and social factors (De Bruyn et al. 1998).

Land degradation (LD) resulting in desertification risk is a multifaceted phenomenon that diminishes soil fertility and initiates ecological and economic processes that define the environment across various spatial scales (Blaikie and Brookfield 1987). LD in arid regions can evolve into an irreversible phenomenon of environmental decline known as desertification (UN 1977). LD is regarded as a significant global threat due to its association with long-term declines in soil productivity and diminished environmental sustainability (Arshad and Martin 2002). LD in arid and semi-arid zones has been a recurring phenomenon throughout human history (UN 1977). Historically, occurrences were sporadic and confined to specific regions, influenced by particular combinations of natural and social factors.

The current system of production, finance, trade and consumption has rapidly expanded from Western Europe since the sixteenth century, leading to the widespread occurrence of LD in arid regions globally (García-Latorre et al. 2001). This cannot be adequately accounted for by recent evidence of climatic changes (e.g. Enne and Zucca 2000). Global socio-economic processes, along with local factors, serve as driving forces for LD (Herrmann and Hutchinson 2005). LD has typically been linked to geo-physical conditions such as soil characteristics, slope and vegetation cover, alongside drought features (e.g. Incerti et al., 2007). However, in Mediterranean regions, a multidisciplinary approach that analyses the interplay of social, economic and ecological factors is essential for accurately identifying sensitive areas (Brouwer and Chadwick 1991).

This article examines significant social dimensions of LD and desertification, addressing issues pertinent to integrated socio-economic and environmental sustainable development. The influence of specific social variables on LD processes is examined, with a focus on the potential correlation between poverty and LD in the Mediterranean basin, illustrated through key examples from rural areas in southern Italy. This study aims to define the desertification process while considering all relevant social dimensions. Focus will be placed on (i) the political, demographic, and economic processes that generate, sustain, and influence desertification, (ii) the human and social factors and outcomes and (iii) the effects of the phenomenon on human organization and existence.

2. Regional Drivers of LD

The causes of LD can be categorized into two types: natural and anthropogenic. The history of LD in the Mediterranean aligns with the evolution of these two groups of causes and their interactions (García-Latorre et al. 2001). Typical causes increasing LD sensitivity include aridity, poor soil characteristics, and the concentration of population and economic activity in coastal areas due to urban growth, internal migration, tourism and intensive agriculture, all closely linked to the unsuitable exploitation of water resources. The crisis in traditional agricultural systems, characterized by land abandonment by rural populations and the resulting deterioration of eroding soil in sloping areas, constitutes a significant issue in marginal rural regions, particularly in northern Mediterranean countries (Puigdefábregas and Mendizabal 1998).

From an ecological perspective, climate, vegetation and soil characteristics are the primary factors influencing land vulnerability to degradation (Salvati et al. 2005). A multivariate analysis conducted to identify ecological indicators quantifying LD sensitivity in Italy at a detailed geographical scale indicates that climate aridity and soil quality are the primary variables accounting for spatial patterns of LD, explaining the highest proportion of variance (Salvati et al. 2006). Aridity effectively illustrates the impact of climatic changes on LD, as it encompasses both the decrease in rainfall and the rise in temperature, which are the most pronounced effects of alterations in climate patterns. The physical processes associated with LD are linked to soil characteristics and the rate of erosion. The impact of land use and cover changes on LD is attributed to practices including agricultural intensification in coastal regions, deforestation in lowland and hilly areas, abandonment of farms and grazing in marginal interior lands and unsustainable woodland management in mountainous regions. Deforestation results in heightened overland flow due to the removal of vegetation, which significantly influences runoff rates, more than any other single factor. This alteration disrupts the long-term social equilibrium in local communities reliant on agriculture and forest economies (e.g. Cammeraat 1995; Cammeraat et al. 2003).

A simplified model illustrating the factors that may accelerate LD is presented here (Figure 1), based on the principal findings from recent studies on this topic (e.g. Rubio and Bochet 1998, Knerr 1998, Enne and Zucca 2000, Cuffaro 2001, García-Latorre et al. 2001, Tanrivermis 2003, Salvati and Zitti 2005). Demographic change and economic development in coastal

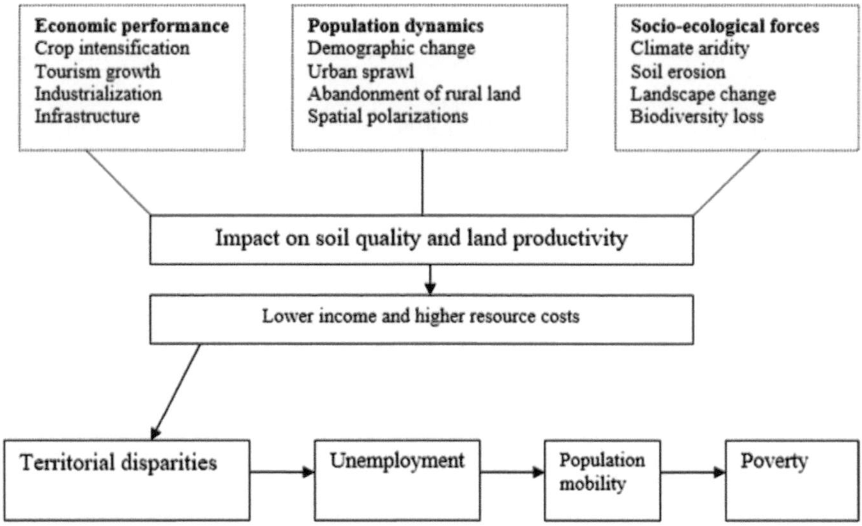

Figure 1. Factors leading to desertification risk and social consequences of poverty.
Source: Authors' elaboration of recent literature.

areas significantly impact urban growth and sprawl, internal migration and population concentration in these regions, often linked to the inappropriate exploitation of natural resources. Furthermore, the decline in population and the aging of farmers contribute to a crisis in traditional and marginal rural systems. Consequent land abandonment results in soil and water erosion, particularly in steeper regions.

3. Soil Degradation and the Socio-ecological Context

3.1. *Population dynamics*

Over the last 50 years, human pressure has intensified, becoming a primary driver of environmental degradation. A common hypothesis suggests that population movements create a mismatch between a region's economic carrying capacity and its population density in areas affected by LD. 'Carrying capacity' refers to an economy's ability to support a certain population size based on its physical capital, human resources, technology, and trade opportunities. Ecologically, it's the maximum number of people an area can sustainably support without degrading their QoL (Blaikie and Brookfield 1987; Arrow et al. 1995). Typically, as population density grows, a society will find ways to increase its carrying capacity, for example, through technological innovation. If it cannot, people may choose to emigrate. These movements can create persistent instability, endangering both the natural environment and its inhabitants.

However, there is no simple, direct causal link between high population numbers and LD. The relationship is more complex. The combination of land sensitivity, population growth rates and other critical factors – like land use patterns, social conditions, and farming practices – is what truly matters. While not a direct cause, population pressure can act as a catalyst, intensifying the severity of LD when combined with other physical, social and economic factors.

3.2. *Increasing urbanization and tourism pressure*

Population growth directly contributes to soil sealing, as human settlements expand over productive land. This urbanization is closely linked to the degradation of high-quality soils, an increased risk of fires, the loss of natural vegetation and the salinization of groundwater.

The influx of tourism further accelerates land consumption, harming fragile natural landscapes and leading to the over-extraction of water resources. Unplanned urban expansion to accommodate tourism has often resulted in inadequate essential services, such as a reliable water supply. However, some research indicates that well-managed tourism can also contribute to conservation efforts in specific contexts (Loumou et al. 2000).

3.3. *Intensification of agriculture*

In economic terms, agricultural intensity refers to the amount of capital, labor, and skill applied to a fixed area of land (Turner and Doolittle 1978). Intensification usually involves expanding cropped areas, increasing cultivation frequency and using more advanced techniques.

However, this intensification is a major contributor to soil and habitat degradation. Practices like deep drainage, large-scale irrigation, heavy pesticide use and multiple cropping are damaging agricultural and semi-natural habitats across vast regions (Venezian-Scarascia et al. 2006). Furthermore, human activities like overstocking, over-cultivation, and deforestation are known to directly fuel desertification. These activities are themselves shaped by socio-economic factors, including land property rights, governing institutions, cultural traditions, and demographic trends.

4. Social Impacts of LD

LD sets off a cascade of severe socio-economic consequences, including the displacement of populations and forced migration. In arid regions, this process can compel people to leave unproductive lands in search of better incomes in urban and coastal areas. This often creates a punishing feedback loop where human behaviours are both a cause and a result of LD, making the problem difficult to address with effective policies. Furthermore, as resources become scarcer, LD can ignite political conflicts over environmental rights.

4.1. *Reduction of agricultural productivity*

A primary economic impact of LD is the decline in agricultural output. The naturally low productivity of dry zones, coupled with unpredictable rainfall, has historically discouraged investment. While policies that prioritize funding for more fertile regions may seem economically rational, they can trigger a downward spiral in arid areas. This lack of investment fosters poor land management and weak economic performance, which is only worsened by the ongoing degradation of natural resources. Consequently, a significant disparity emerges within the rural sector between 'favourable' and 'less favourable' lands, placing immense pressure on local communities.

4.2. *Migration and depopulation of rural areas*

Environmental degradation is intrinsically linked to complex patterns of human migration. These movements can be seasonal or permanent and can occur between rural areas, from rural to urban centres or internationally. Desertification, in particular, is a notable driver of migration, as communities respond to the growing imbalance between the land's carrying capacity and population density. As the environmental resource base erodes, the number of people unable to maintain their living standards is likely to rise, which can propagate poverty and social conflict over resources like water and land (Chopra and Gulati 1997; Boyce 1994).

The direct consequence of this migration is the depopulation of rural and marginal areas. Farmers are often the first victims of desertification, as their livelihoods are immediately impacted (Chopra and Gulati 1997; Pender 1998). The abandonment of farmland, a trend that began in Southern Europe in the 1950s due to industrialization and shifting economic policies, is a key outcome of LD combined with these population shifts. The future of this abandoned land depends on its condition at the time of abandonment; subsequent activities like intensive grazing can accelerate soil erosion, especially on sloped terrain (e.g. UN 1977).

4.3. *Unemployment and loss of rural culture*

The concentration of populations in coastal and lowland areas can lead to increased unemployment, which disproportionately affects vulnerable groups like women and youth. Local labour markets tied to both the primary sector and tourism are sensitive to the production declines and environmental decay caused by LD. Alongside these economic impacts is the erosion of culture. Sustainable land management relies on traditional knowledge and skills. With little large-scale information available from farmers on combating desertification, it is crucial to encourage community exchanges to prevent the loss of this invaluable knowledge.

4.4. *Poverty and emerging resource conflicts*

Poverty is widely considered both a cause and a consequence of LD, and statistics from arid regions show a strong geographic correlation between poverty, drought and desertification. The relationship is not simple; rather, poverty often serves as a mechanism through which other factors, such as inadequate policies or institutions, drive degradation. The literature describes this link in three main ways: the overly simplistic 'Vicious Cycle' model; a more nuanced 'Indirect Link' that accounts for external factors; and the 'Amplifying Role of Poverty', where poverty worsens degradation in the absence of strong public support. Impoverished populations in these regions are often geographically isolated, vulnerable and lack access to essential public goods. Many are pushed into ecologically fragile areas by socio-economic pressures like land expropriation and the privatization of common lands (Chopra and Gulati 1997).

This pressure on fragile lands exacerbates competition for scarce natural resources like land, water and energy, leading to conflicts that can widen regional inequalities. Conflicts over water are especially common in drought-prone developed countries like Spain, Italy and Greece, creating significant barriers to sustainable development.

4.5. *The downward spiral and territorial dichotomy*

The combination of these factors can trap communities in a 'poverty-LD' downward spiral. This concept, often called the 'poverty trap', describes a cycle where people in poverty must degrade the natural resources they depend on to survive, making them both agents and victims of environmental erosion. In the Mediterranean, this spiral manifests as a stark territorial dichotomy, with a growing divergence between developing coastal municipalities and declining inland areas (e.g. Kosmas et al. 1999). This division between the interior and the coast in terms of services, opportunities, and human potential appears to be a significant cause of LD itself. The dynamics of this relationship are shaped by property rights, governing institutions, and demographic traditions (Enne and Zucca 2000). A conceptual framework illustrating these divergent trends over time is presented in Figure 2. The model plots the change in 'Poverty Risk' against 'Desertification Risk' for different types of districts from a starting time (t_0) to a later time (t_1). It suggests that:

> Lowlands and accessible coastal areas experience a significant increase in desertification risk, driven by ecological pressures, but a relatively small increase in poverty risk.

In contrast, Mountainous and peripheral rural districts show a large increase in poverty risk, linked to social pressures like depopulation and resource depletion, while their desertification risk increases less dramatically. Uplands and intermediate districts follow a path between these two extremes. This framework visually demonstrates how LD and poverty evolve differently across a territory, exacerbating the social and economic disparities between coastal and inland regions.

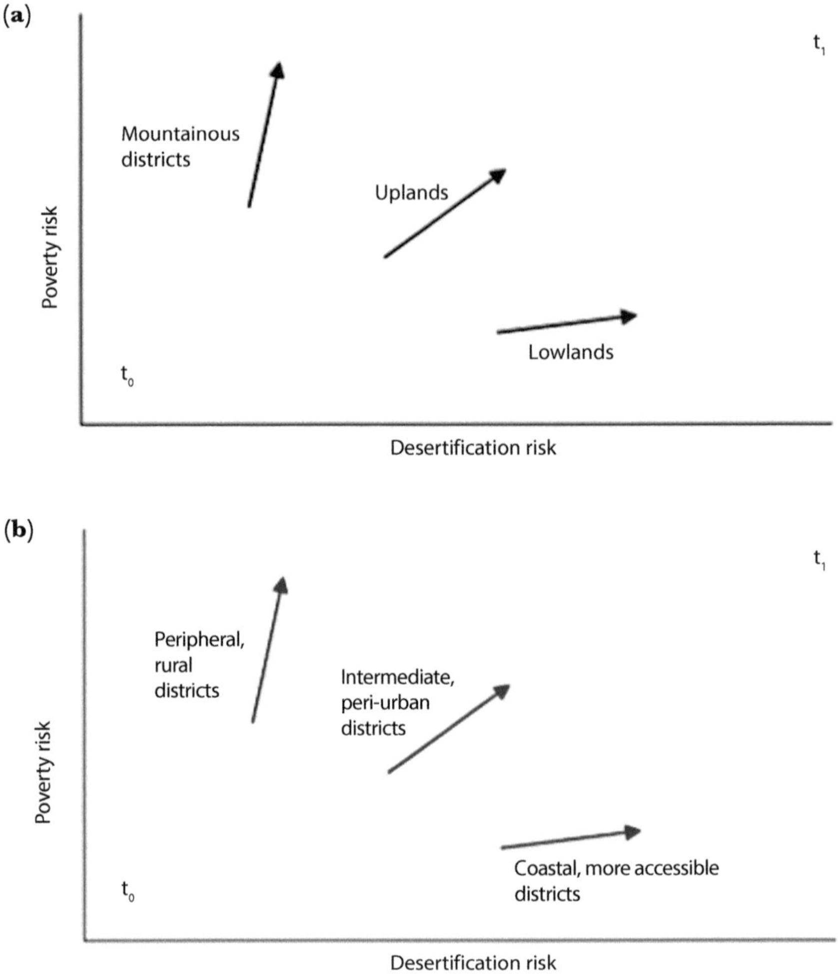

Figure 2. Trends over time in desertification and poverty risk in the Mediterranean context along (a) the elevation gradient and (b) the urban-rural gradient.
Source: Authors' elaboration of recent literature.

4.6. *Land resource management*

Effective land resource management is influenced by a complex mix of political, economic and social factors. The structure of land ownership is critically important. While it was once believed that communal ownership inevitably leads to degradation, research shows that communal resources can be managed effectively with active participation and strong local institutions. Conversely, in frameworks that promote land concentration, private ownership and unregulated land markets can lead to speculation and short-term profit-seeking that harms the environment. A key factor in improving resource management is establishing permanent mechanisms for environmental education and providing economic incentives for sustainable practices, such as supporting the transition to organic farming in marginal areas.

5. Assessing the Relationship Between LD and Poverty: Social Indicators

The complexity of cause–effect relationships and the linkage between LD and human activities make it difficult to find efficient methodologies to assess the interrelationships between environmental and social factors (Rubio and Bochet 1998). A significant challenge noted in the most recent literature on LD is the considerable lack of time series and scenario analyses on vulnerable areas and the evolution of the main human factors involved. An integrated evaluation of the social aspects of land configuration to assess LD sensitivity should be carried out through a multi-temporal approach, taking into account the structural changes that occurred during the last fifty years in southern Europe.

In practice, frameworks used to build indexes of LD sensitivity, such as the Environmental Sensitive Area (ESA) procedure, have generally considered few social variables (e.g. Basso et al. 2000). Examining the Italian experience in terms of assessment of desertification risk, the most wide-ranging effort to produce reliable estimations of the surface at risk occurred in the last ten years and produced more than twenty maps at both national and regional levels (Ceccarelli et al. 2006). Among these works, indicators such as population density and demographic variation over a decade were considered in only two maps (both at a national level), while urban sprawling, elderly index, illiteracy index, retirement index, as well as an employment index were included in only one map (at the regional level). To improve this, a panel of social and economic indicators should be promptly used in the cartographic evaluation of the desertification risk through standard procedures (e.g. Enne and Zucca 2000, Cammeraat et al. 2003). The candidate indicators should be: (i) easily computed at the highest spatial resolution possible, (ii) available at least every ten years and (iii) recognizable as key descriptors in the fields of demography, agriculture, industry, tourism, and the labour market. Based on these criteria and previous contributions on the topic (e.g. Rubio and Bochet 1998, Cammeraat et al. 2003, Salvati and Zitti 2005, Ceccarelli et al. 2006, Salvati and Zitti 2007), Table 1 presents a restricted panel of indicators able to depict the social factors of LD in Mediterranean Europe.

Ultimately, giving precise spatial information on the speed of LD and on the variation in its human causes may represent a key tool to drive political

Table 1. Examples of social indicators able to depict human factors involved in LD processes.

Social	Demographic	Economic	Institutional/Policy
Poverty Gap Index	Population density	Gross domestic product	Scientific research funding
Unemployment rate	Demographic variation	Tourism density	Regional/local funding
GDP spent in education	Total fertility rate	Industrial concentration	Agri-environmental measures
Crime rate	Net migration rate	Agricultural intensity	Water use policy
Literacy index	Ageing rate	Farm income	Protected areas
Public perception of LD	Urban sprawl rate	Farm subsidies	Urban planning
	Farmer ageing	Crop productivity	Best practices

Source: Authors' elaboration of recent literature.

actions able to effectively mitigate processes of desertification at a local scale (Ceccarelli et al. 2006). This requires a concerted effort to analyse sociological and statistical methods, establish data warehouses with tools to analyse complex datasets, define the correct geographical scale to recognize social and economic indicators and prepare manuals of good practices for authorities.

6. An Example of Downward Spiral: The Case of Italy

At the European scale, anticipated land use changes and their implications for sustainability and soil resources have been systematically examined (Bouma et al. 1998). In Italy, LD phenomena are prominently observed in the southern regions, though northern areas have also shown increased sensitivity due to drought and intensive agriculture (Salvati et al. 2005, 2006). The country provides a compelling case study of how environmental degradation and socio-economic factors interact.

The spatial distribution of this issue is illustrated in Figure 3, which maps LD Sensitivity. The map reveals that while southern regions like Apulia, Sardinia and Sicily show high sensitivity – where this degradation is termed desertification – other significant sensitive areas exist, notably in the northern Po Valley, linked to intensive agricultural practices. This clustered geography reflects the importance of spatial dependence in regional dynamics. Economic and environmental vulnerabilities in one area often reinforce those of neighbouring regions (Arbia 2001). This spatial logic underscores why Italy's well-documented North–South divide persists and why localized degradation risks cannot be fully understood without considering regional interconnections. Moreover, these dynamics mirror historical population trends, showing growth in coastal and lowland regions marked by high degradation sensitivity, and population decline in the less-sensitive mountain zones.

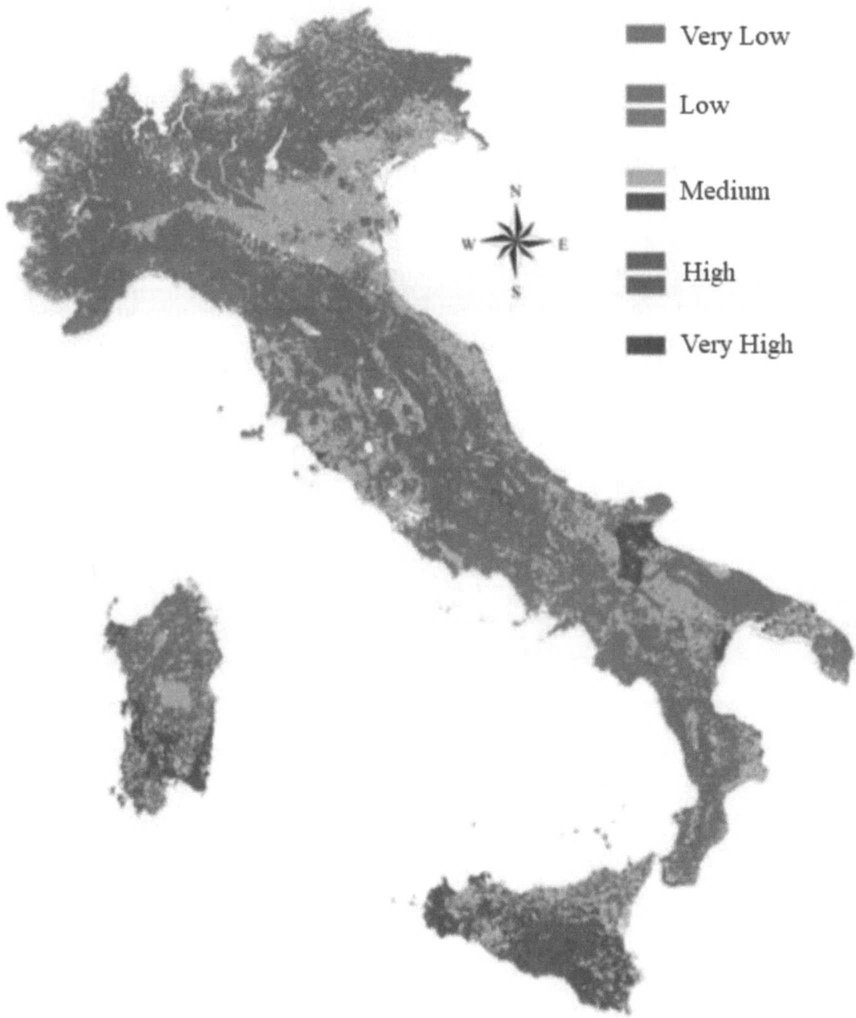

Figure 3. Land Degradation Sensitivity in Italy.
Source: Adapted from Boschetto, Di Leginio and Luise (2010, 22).

This environmental context is deeply connected to Italy's well-documented North-South economic divide (Bagnasco 1977; Trigilia 2005). The connection between these territorial disparities and environmental risk is quantified in Figure 4. The scatter plot demonstrates a clear positive correlation between the 'Desertification risk index' and the percentage of the 'Population under the poverty line' at a regional level. The upward slope of the trend line provides strong statistical support for the 'poverty-LD' downward spiral hypothesis; as environmental vulnerability increases, so does the prevalence of poverty.

To understand the specific mechanisms driving this spiral, a more detailed framework may focus on the agricultural economy of Southern Italy, illustrating how factors like increasing agricultural intensity and mechanization lead to greater water extraction from wells. This, combined with climatic changes and unsustainable water management, causes soil salinization, which in turn reduces soil fertility and agricultural productivity. This decline in the primary sector, which is disproportionately important to the southern economy, directly contributes to rising poverty.

Finally, the real-world social consequences of this process are confirmed by the empirical data in Table 2. The table classifies municipalities in Southern Italy by their level of soil degradation and compares key social indicators. The data consistently show worse social conditions in areas with higher

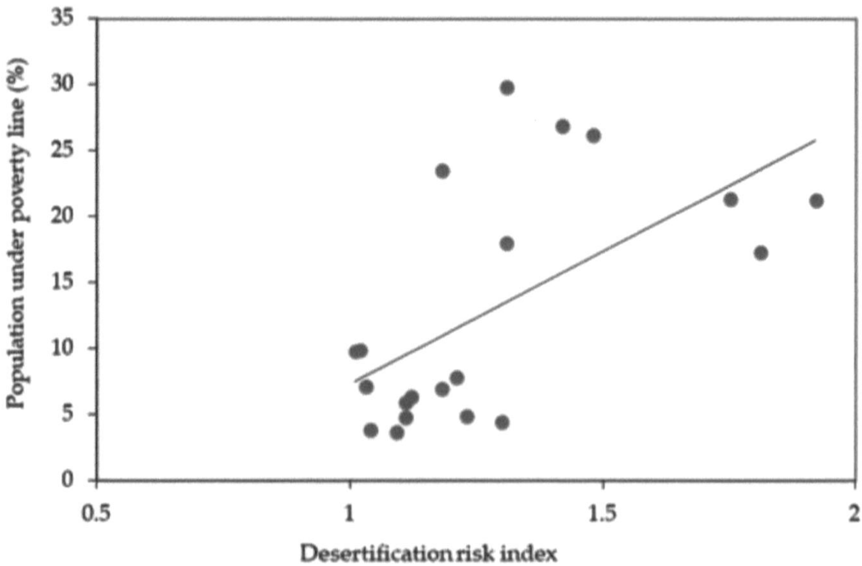

Figure 4. Relationship between desertification risk and relative poverty in Italy by administrative region.

Source: Authors' elaboration of recent literature.

Table 2. The latent relationship between soil degradation and selected indicators in Italy.

Variable	*Soil degradation level*				*Southern Italy*	Italy
	Absent	Low	Medium	High		
Unemployment index (total)	18.6	18.8	20.0	23.3	*20.8*	**11.9**
Unemployment index (young)	48.1	48.9	49.2	54.4	*50.9*	**30.7**
Unemployment index (females)	24.7	25.5	27.5	32.0	*28.5*	**16.6**
Unemployment index (young females)	55.1	57.0	58.2	65.1	*60.2*	**37.1**
Employees in the primary sector (%)	12.0	14.0	16.0	14.0	*15.0*	**10.0**
Temporary employees in nonprimary sector (%)	2.1	3.7	3.4	3.7	*3.6*	**4.0**
Population dropping out primary education (5)	11.8	12.3	13.7	14.4	*13.6*	**11.1**
Population with tertiary education degree (%)	4.6	5.1	5.0	4.9	*5.0*	**5.0**
Elderly index of farmers (% >55 years)	18.0	20.0	21.0	21.0	*21.0*	**23.0**
Farmers with technical degree (%)	18.0	17.0	16.0	17.0	*17.0*	**19.0**
Crime rate (per 1.000 inhabitants)	14.6	13.0	17.6	24.3	*18.8*	**20.4**

Source: Authors' elaboration of recent literature.

degradation. For example, unemployment rates – especially for women and youth – are dramatically higher in the most degraded areas, while education levels show an inverse pattern. The table also highlights the significant issue of an aging farmer population in these vulnerable zones. These data provide concrete proof of the latent relationship between environmental health and social well-being.

7. Modern Context: Climate Change, Policy Frameworks and New Solutions

Effective strategies to monitor and combat desertification have been hindered by fragmented frameworks, weak institutional integration and lack of systematic monitoring – obstacles highlighted in global analyses of desertification control failures (Glenn et al. 1998). The challenges of LD and poverty are increasingly amplified by climate change. The Mediterranean region is a recognized climate change hotspot, warming 20% faster than the global average (MedECC 2020). This accelerated warming intensifies the drivers of desertification, leading to more frequent and severe droughts, which directly threaten water availability and agricultural viability (MedECC 2020). This climatic pressure interacts with and exacerbates the socio-economic vulnerabilities discussed throughout this chapter.

In response, international and regional policy frameworks have become more integrated. The United Nations' 2030 Agenda for Sustainable Development provides a global blueprint for addressing these interconnected issues. Specifically, Sustainable Development Goal (SDG) 15 (Life on Land) calls for action to 'combat desertification, restore degraded land and soil […] and strive to achieve a land degradation-neutral world' by 2030 (UN 2015). This goal is intrinsically linked to SDG 1 (No Poverty), as LD disproportionately affects the livelihoods of the rural poor (UN 2015; UNCCD 2017). The UN recognizes that protecting and restoring land is a fundamental driver of economic growth and poverty reduction.

At the European level, the European Green Deal and its associated EU Soil Strategy for 2030 represent a significant policy evolution. The strategy acknowledges that around 60–70% of EU soils are estimated to be in an unhealthy state (European Commission 2021). It sets a vision for all EU soil ecosystems to be healthy by 2050 and proposes concrete actions, including the development of a dedicated legislative proposal on soil health, promoting sustainable soil management and developing a common methodology to assess desertification (European Commission 2021). These policies aim to shift from a fragmented, sectoral approach to a holistic one that integrates soil health into broader climate, biodiversity and agricultural frameworks.

Alongside policy, there is growing interest in Nature-Based Solutions (NbS), which are 'actions to protect, sustainably manage, and restore natural or modified ecosystems, that address societal challenges effectively and adaptively, simultaneously providing human well-being and biodiversity benefits' (Cohen-Shacham et al. 2016). In the Mediterranean, NbS projects include the restoration of wetlands and floodplains to mitigate floods and droughts, the use of cover crops and agroforestry to improve soil health, and the re-establishment of riparian buffers to prevent erosion. These approaches can offer cost-effective and adaptable alternatives to purely technological interventions, building long-term resilience (Cohen-Shacham et al. 2016).

8. Concluding Remarks

LD and desertification are, at their core, social issues. The ways we use and manage land are socially constructed, meaning that concepts like productivity and sustainability emerge from the interaction between people and nature. The fundamental causes of desertification are deeply connected to human activities, environmental pressures, social and economic structures and unsustainable practices in resource production and consumption (Blaikie and Brookfield 1987; Antle and Heidebrink 1995). Overexploitation of resources, deforestation and uncontrolled urban growth are all symptoms of these deeper issues.

Therefore, social policies are essential for tackling LD. A comprehensive approach, aligned with modern frameworks like the SDGs and the EU Green Deal, should include policies that:

- Combat poverty and inequality, especially in rural and marginalized areas, to create a framework where people can adopt sustainable practices.
- Ensure gender equality, as poverty is often a gendered issue. Empowering women in economic and social life is critical.
- Prioritize human resource development through education and training with a strong focus on environmental sustainability.
- Implement effective migration policies that address population pressures and promote sustainable development in both rural and urban areas.
- Promote and fund NbS and sustainable land management practices that build resilience to climate change (Arnalds 2000).

The connection between social factors and LD must be discussed at all levels – from local communities to national governments – to develop effective policies that mitigate the impacts of drought and desertification, particularly for Southern European countries. Addressing desertification is not just an environmental issue; it is fundamental to achieving sustainable development and securing the long-term well-being of people in arid regions.

References

Antle, John M., and Gregg Heidebrink. 1995. "Environment and Development: Theory and International Evidence." *Economic Development and Cultural Change* 43(3): 603–25. https://doi.org/10.1086/452171.

Arbia, Giuseppe. 2001. The role of spatial effects in the empirical analysis of regional concentration. *Journal of Geographical Systems* 3, 271–281. https://doi.org/10.1007/PL00011480.

Arnalds, Olafur. 2000. "Desertification: An Appeal for a Broader Perspective." In *Rangeland Desertification*, edited by Olafur Arnalds and Steven Archer, 5–15. Dordrecht: Springer. https://doi.org/10.1007/978-94-015-9602-2_2.

Arrow, Kenneth, Bert Bolin, Robert Costanza, Partha Dasgupta, Carl Folke, Crawford S. Holling, Bengt-Owe Jansson, et al. 1995. "Economic Growth, Carrying Capacity and the Environment." *Ecological Economics* 15 (2): 91–95. https://doi.org/10.1016/0921-8009(95)00059-3

Arshad, Muhammad A. Charlie, and S. Martin. 2002. "Identifying critical limits for soil quality indicators in agro-ecosystems." *Agriculture, Ecosystems & Environment* 88(2): 153–160. https://doi.org/10.1016/0921-8009(95)00059-3.

Bagnasco, Arnaldo. 1977. *Tre Italie. La problematica territoriale dello sviluppo italiano.* IL Mulino, Bologna.

Basso, Francesco, Ettore Bove, Stefano Dumontet, Agostino Maria Silvio Ferrara, Michele Pisante, Giovanni Quaranta, and Malcolm J. Taberner 2000. Evaluating environmental sensitivity at the basin scale through the use of geographic information systems and remotely sensed data: an example covering the Agri basin – Southern Italy. *Catena* 40: 19–35. https://doi.org/10.1016/S0341-8162(99)00062-4.

Blaikie, Piers, and Harold Brookfield. 1987. *Land Degradation and Society* (1st ed.). Routledge. https://doi.org/10.4324/9781315685366.

Boschetto, Riccardo G., Marco Di Leginio, and Anna Luise. 2010. An Example of a Country Assessment: Desertification Mapping in Italy. *Italian Journal of Agronomy*. 5(3): 21–26. https://doi.org/10.4081/ija.2010.s3.21.

Bouma, Johan, György Várallyay, and Niels H. Batjes. 1998. Principal land use changes anticipated in Europe. *Agriculture, Ecosystems and Environment, 67*, 103–119. https://doi.org/10.1016/S0167-8809(97)00109-6.

Boyce, James K. 1994. Inequality as a cause of environmental degradation. *Ecological Economics* 11(3): 169–178. https://doi.org/10.1016/0921-8009(94)90198-8.

Brouwer, Floor M., and Martin J. Chadwick. 1991. "Future Land Use Patterns in Europe." In *Land Use Changes in Europe*, edited by F. M. Brouwer, A. J. Thomas, and M. J. Chadwick, 49–78. GeoJournal Library, vol. 18. Dordrecht: Springer. https://doi.org/10.1007/978-94-011-3290-9_3.

Cammeraat, Erik. 1995. The MADALUS Core Field Programme: An Overview of Sites and Methodology. In *Mediterranean desertification and land use*, edited by Thornes J.B. & Jane Brandt, 87–110. John Wiley & Sons, Chichester, UK.

Cammeraat, Erik, Anton Imeson, and Lars Hein. 2003. Investigation on environmental characteristics to underpin the selection of desertification indicators in the Guadalentin basin. In *Mediterranean desertification: a mosaic of processes and responses*, edited by Geeson, Nichola A., C. Jane Brandt, and John B. Thornes, 187–203. John Wiley & Sons: Chichester, UK.

Ceccarelli, Tomaso, Francesca Giordano, Anna Luise, Luigi Perini, and Luca Salvati. 2006. *La vulnerabilità alla desertificazione in Italia: raccolta, analisi, confronto e verifica delle procedure cartografiche di mappatura e degli indicatori a scala nazionale e locale.* Manuali e Linee Guida no. 40, APAT – CRA UCEA. Rome, Italy.

Chopra Kanchan, and Gulati Sahil C. 1997. Environmental degradation and population movements: the role of property rights. *Environmental and Resource Economics*. 9: 383–408. https://doi.org/10.1007/BF02441758.

Cohen-Shacham, Emmanuelle, Gretchen Walters, Christine Janzen, and Stewart Maginnis. 2016. *Nature-based Solutions to Address Global Societal Challenges*. Gland, Switzerland: IUCN. https://doi.org/10.2305/IUCN.CH.2016.13.en.

Cuffaro, Nadia. 2001. *Population, Economic Growth and Agriculture in Less Developed Countries* (1st ed.). Routledge. https://doi.org/10.4324/9780203103104.

De Bruyn, Sander M., Jeroen C.J.M van den Bergh, and Johannes Baptist Opschoor. 1998. Economic growth and emissions: reconsidering the empirical basis of environmental Kuznets curves, *Ecological Economics*, 25(2): 161–175. https://doi.org/10.1016/S0921-8009(97)00178-X.

Enne, Giuseppe, and Claudio Zucca. 2000. *Desertification indicators for the European Mediterranean region: state of the art and possible methodologicalapproaches [= Indicatori di desertificazione per il Mediterraneo europeo: stato dell'arte e proposte di metodo]*. Agenzia Nazionale per la Protezione dell'Ambiente: Rome, Italy.

European Commission. 2021. *Communication from the Commission to the European Parliament, the Council, the European Economic and Social Committee and the Committee of the Regions: EU Soil Strategy for 2030 – Reaping the Benefits of Healthy Soils for People, Food, Nature and Climate.* Brussels: European Commission. Accessed August 2, 2025. https://eur-lex.europa.eu/legal-content/EN/TXT/?uri=CELEX:52021DC0699.

García-Latorre, Jesús, Juan Garcı́a-Latorre and Andrés Sanchez-Picón. 2001. "Dealing with aridity: socio-economic structures and environmental changes in an arid Mediterranean region." *Land Use Policy* 18: 53–64.

Glenn, Edward, Mark Stafford Smith, and Victor R. Squires. 1998. On our failure to control desertification: implications for global change issues, and research agenda for the future. *Environmental Science & Policy* 1: 71–78. https://doi.org/10.1016/S1462-9011(98)00007-0.

Herrmann, Stefanie M., and Charles F. Hutchinson. 2005. The changing contexts of the desertification debate. *Journal of Arid Environments* 63(3): 538–555. https://doi.org/10.1016/j.jaridenv.2005.03.003.

Hussen, Ahmed. 2004. *Principles of Environmental Economics: An Integrated Economic and Ecological Approach* (2nd ed.). Routledge. https://doi.org/10.4324/9780203507421.

Incerti, Guido, Enrico Feoli, Luca Salvati, Antonio Brunetti, and Aldo Giovacchini. 2007. Analysis of bioclimatic time series and their neural network-based classification to characterise drought risk patterns in South Italy. *International Journal of Biometeorology*, 51:253–263. https://doi.org/10.1007/s00484-006-0071-6

Knerr, Béatrice. 1998. "Impacts of Labour Migration on the Sustainability of Agricultural Development in Arid Regions." In *Population and Environment in Arid Regions*, edited by J. Clarke and D. Noin, 211–246. Man and the Biosphere Series 19. New York: UNESCO.

Kosmas, Costas, Mike Kirkby, and Geeson, N. 1999. *The MEDALUS project: Mediterranean desertification and land use: Manual on key indicators of Desertification and mapping environmental sensitive areas to desertification.* Luxembourg: European Commission – Project Report. https://catalogue.unccd.int/629_Medalus_ESA_%20Manual.pdf.

Loumou Aggela, Giourga Christina., Panayiotis Dimitrakopoulos, and Sotirios Koukoulas. 2000. PROFILE: Tourism Contribution to Agro-Ecosystems Conservation: The Case of Lesbos Island, Greece. *Environmental Management* 26(4): 363–370. doi:10.1007/s002670010093.

MedECC (Mediterranean Experts on Climate and Environmental Change). 2020. *Climate and Environmental Change in the Mediterranean Basin – Current Situation and Risks for the future.* First Mediterranean Assessment Report. Marseille: Plan Bleu. https://www.medecc.org/wp-content/uploads/2021/05/MedECC_MAR1_complete.pdf

Nocenzi, Mariella. 2005. "Environment and Value Challenge: Persistencies and Changes." *International Review of Sociology* 15 (2): 291–92. doi:10.1080/03906700500159623.

Pender, John L. 1998. Population growth, agricultural intensification, induced innovation and natural resource sustainability: an application of neoclassical growth theory. *Agricultural Economics* 19: 99–112. https://doi.org/10.1016/S0169-5150(98)00024-3.

Puigdefábregas, Juan, and Teresa Mendizabal. 1998. Perspectives on desertification: western Mediterranean. *Journal of Arid Environments*, *39*(2): 209–224. https://doi.org/10.1006/jare.1998.0401.

Rubio José Luís, and Esther Bochet. 1998. Desertification indicators as diagnosis criteria for desertification risk assessment in Europe. *Journal of Arid Environments*. 39(2): 113–120. https://doi.org/10.1006/jare.1998.0402.

Salvati, Luca, and Marco Zitti. 2005. Land degradation in the Mediterranean Basin: Linking bio-physical and economic factors into an ecological perspective. *Biota*, 6: 67–77. EID: 2-s2.0-33747486236.

———. 2007. Long term demographic dynamics along an urban-rural gradient: implications for land degradation. *Biota* 8(1):61–69. EID: 2-s2.0-54149111383.

Salvati, Luca, Tomaso Ceccarelli, and Antonio Brunetti. 2005. "Desertificazione, cambiamenti climatici e agricoltura in Italia: Primi risultati di un modello di valutazione del rischio di desertificazione." *AgriStat–Statistiche Agricole*. 2005.

Tanrıvermiş, Harun. 2003. Agricultural land use change and sustainable use of land resources in the Mediterranean region of Turkey. *Journal of Arid Environments* 54: 553–564. https://doi.org/10.1006/jare.2002.1078.

Trigilia, Carlo. 2005. *Sviluppo locale. Un progetto per l'Italia*. Saggi Tascabili Laterza, Roma-Bari.

Turner, Billie Lee II, and William E. Doolittle. 1978. "The concept and measure of agricultural intensity." *The Professional Geographer* 30(3): 297–301. doi: 10.1111/j.0033-0124.1978.00297.x.

UN (United Nations). 1977. *Report of the United Nations Conference on Desertification*. New York: United Nations. https://digitallibrary.un.org/record/733000?v=pdf.

UN (United Nations). 2015. *Transforming Our World: The 2030 Agenda for Sustainable Development*. New York, United Nations. .

UNCCD (United Nations Convention to Combat Desertification). 2017. *The Global Land Outlook*. UNCCD. Bonn, Germany. https://digitallibrary.un.org/record/3972346?v=pdf.

Venezian-Scarascia, Maria Elisa, Federica Di Battista, and Luca Salvati. 2006. Water resources in Italy: availability and agricultural uses. *Irrigation and Drainage*. 55: 115–127. https://doi.org/10.1002/ird.222.

Chapter 6

LOCAL WELLBEING, OPTIMAL LAND MANAGEMENT AND SPATIAL DIVIDES IN WASTE INFRASTRUCTURE: A PRE-CRISIS PERSPECTIVE FOR ITALY

Bünyamin Özkür, Gianluca Egidi, Luigi Fici, Maurizio Masi, Ioannis Konaxis, Donato Scarpitta, Marco Maialetti and Adele Sateriano

1. Introduction

Modern economic infrastructure provides the essential backbone for social well-being and economic growth; its presence in a region enhances QoL and spurs development (UN 2015). Within this framework, waste management systems have traditionally been viewed through a narrow, technical lens: a utility service designed to dispose of the unwanted by-products of consumption (Wilson 2007). This perspective, however, is dangerously obsolete. A contemporary understanding, compelled by the planetary crises of climate change, resource depletion and biodiversity loss, demands a paradigm shift (Wilson 2015). This chapter argues for a conceptual evolution: from viewing 'waste management' as a disposal problem to understanding it as a central pillar of resource governance, a domain inextricably linked to national economic strategy, social equity and environmental health (Park et al. 2024; OECD 2013).

Effective waste management is both an outcome and an enabler of sustainable development (UN 2025; UN 2023). Its principles are deeply embedded within the UN's Sustainable Development Goals (SDGs), touching upon the creation of sustainable cities (SDG 11), ensuring responsible consumption and production (SDG 12), reducing inequalities (SDG 10), and protecting public health (SDG 3) (UN 2015; Kaza et al. 2018). The transition from a linear 'take-make-dispose' economy to a Circular Economy (CE) – a system designed to eliminate waste, circulate products at their highest value, and regenerate natural systems – is now recognized as a fundamental strategy for achieving these goals (Ellen MacArthur Foundation n.d.; European Commission 2020; SDG Knowledge Hub 2024). This transition, however, must be a Just Transition, ensuring that the transformation is fair and inclusive, creating decent work opportunities and leaving no one behind (UN-Habitat 2023; ILO 2018). A just transition addresses the social dimensions of economic change, ensuring that the benefits of a greener economy are shared broadly and that vulnerable communities are not disproportionately burdened by the shift (GIZ 2024; IUCN 2024).

This chapter posits that the stark North–South divide in Italy's waste management system, clearly visible in the pre-crisis period of the early 2000s, represents more than a technical or developmental lag. It is a persistent issue of environmental and spatial injustice (Agovino et al. 2016). The infrastructural and policy choices of that era created a path dependency that continues to challenge Italy's ability to achieve a truly sustainable and equitable circular economy (Mazzanti et al. 2006). By treating Italy as a laboratory of distinct spatial areas, this analysis examines how waste management functions as a regional factor that can either bolster or hinder long-term sustainability, revealing latent specializations and disparities that have profound implications for national development (Cerqueti et al. 2021). The persistence of these issues is confirmed by recent data; the ISPRA 2024 report shows that while the national separate collection rate has reached 66.6%, a significant gap remains between the North (73.4%) and the South (58.9%), with the South also lagging in treatment plant infrastructure (ISPRA 2024; Circular Economy Network 2024).

2. Sustainable Development Strategies in Italy: A Policy Review

Italy's formal commitment to sustainable development has a long and complex history, marked by the adoption of ambitious legislative frameworks that often preceded similar actions in other European nations (OECD 2013; Domorenok 2018). As early as 1993, in the wake of the Rio Earth Summit, the country adopted a National Plan for Sustainable Development aligned with the principles of Agenda 21 (MASE 1993; UN 1992). This plan, approved by the influential Interministerial Committee for Economic Planning (CIPE), reflected the international vision articulated in the Brundtland Report (WCED 1987) and signalled Italy's early intent to integrate environmental considerations into sectoral policies and reconcile economic activity with environmental preservation (Domorenok 2018).

This commitment was legally solidified with the passage of Legislative Decree No. 152/2006, widely known as the 'Codice dell'Ambiente' [Environmental Code] (Ambiente Diritto 2006). This landmark legislation, which remains central to Italian environmental law, formally enshrined the principle of sustainable development into national law, defining it as the imperative to meet present needs 'without compromising the quality of life and possibilities of future generations' (Ambiente Diritto 2006). The Code regulates environmental impact assessments, soil and water protection, waste management, and air quality, among other areas. An amendment in 2008 further strengthened this mandate, requiring all public administrations to align their decision-making with this principle of intergenerational equity, explicitly weighing environmental and cultural protection alongside other public interests (ASviS 2018; OECD 2013). This legal codification was a significant step, at least in theory, towards embedding sustainability into the fabric of Italian public policy (ASviS 2023).

These high-level commitments were supported by specific implementation tools, such as the 2008 Action Plan for the Environmental Sustainability of Public Administration Consumption (Piano d'Azione Acquisti Verdi), which leveraged the state's purchasing power to drive markets for greener products (MASE 2008). Following the global adoption of the UN's 2030 Agenda, Italy launched its National Strategy for Sustainable Development (SNSvS) in 2017, localizing the SDGs through a broad consultative process involving ministries, regional governments and civil society (MASE 2017; Domorenok 2018; UN 2015). This strategy was intended to be the umbrella framework for integrating sectoral policies to achieve the SDGs (OECD 2022).

In this context, civil society has played a crucial, proactive, and often critical role. The Italian Alliance for Sustainable Development (ASviS), founded in 2016, emerged as a key actor, uniting over 300 public and private entities, including trade unions, universities, and foundations (ASviS 2021). ASviS works to promote awareness of the SDGs, monitor national progress, and advocate for policies that address territorial gaps and inequalities (ASviS 2021). Through its comprehensive annual reports, ASviS has consistently highlighted the gap between policy ambition and reality, acting as a critical watchdog pressuring the government to translate its legislative ambitions into concrete and equitable outcomes (ASviS 2020). The 2024 ASviS report, for instance, starkly concluded that Italy is on an 'unsustainable development path', with only a fraction of its quantitative targets likely to be met by 2030 (ASviS 2024; Ribellino 2024). This critique points to a persistent failure to implement the National Sustainable Development Strategy and a disconnect between public concern over issues like climate change and the perceived lack of long-term vision in government action (Prioreschi et al. 2023; ASviS 2024). The challenges are compounded by a lack of open data sources and validated metrics at the local level, which hinders effective monitoring and comparison of progress across municipalities (Richiedei and Pezzagno 2022).

3. The Waste Issue as a Signal of Unsustainable Development in Pre-Crisis Italy

3.1. *Overview of waste disposal trends*

Despite Italy's ambitious legislative framework, the reality on the ground during the pre-crisis years of the early 2000s painted a picture of deeply fractured development (Mazzanti et al. 2006). The management of municipal solid waste (MSW) became a powerful indicator of this national dichotomy (Agovino et al. 2016). The period was shaped by the 1997 Ronchi Decree, which first established a compulsory minimum recycling rate and placed responsibility for waste management on municipalities, setting the stage for the divergent paths that would follow (Stodola 2023; Paleari and Ferraris 2013). Figure 1 vividly illustrates this divergence, showing the progress of separate municipal waste collection across Italian provinces from 1999 to 2005. The maps reveal a clear progression in the North, where many provinces (dark brown) surpassed the 40% separate collection threshold by 2005. In stark contrast, the South and the Islands (yellow) showed minimal progress, with most provinces remaining below 20%. This visual evidence points to a severe lag in policy implementation and a deepening spatial divide, reflecting broader inequalities in institutional capacity and public service provision (ASviS 2021; OECD 2013).

During this period, the primary method for MSW disposal across Italy was landfilling, an unsustainable practice at the bottom of the waste hierarchy that poses significant environmental and health risks (Paleari and Ferraris 2013; Wilson 2007). The reliance on this method, however, varied dramatically by region. As shown in Figure 2 (bottom maps), landfilling remained the overwhelmingly dominant practice in the South and the Islands, where in 2005 between 80% and 100% of all disposed waste (darkest purple) was sent to landfills. This occurred even as national and European legislation began to mandate a shift towards treatment and recovery, reserving landfill space only for pre-treated, non-recyclable residual waste (Domorenok 2018; Ambiente Diritto 2006). The environmental consequences of this over-reliance were severe, including the production of leachate that can contaminate soil and groundwater with heavy metals and other toxins, and the emission of landfill gas (methane and CO_2), a major contributor to greenhouse gas emissions (Kalmar ECO-TECH 2012; Stevenazzi et al. 2023). In Campania, this situation escalated into a full-blown crisis, with the illegal dumping and burning of waste creating the so-called 'land of fires' (in Italian, 'Terra dei Fuochi'), leading to documented increases in cancer mortality and other adverse health outcomes in the affected populations (Senior and Mazza 2004; Triassi et al. 2015).

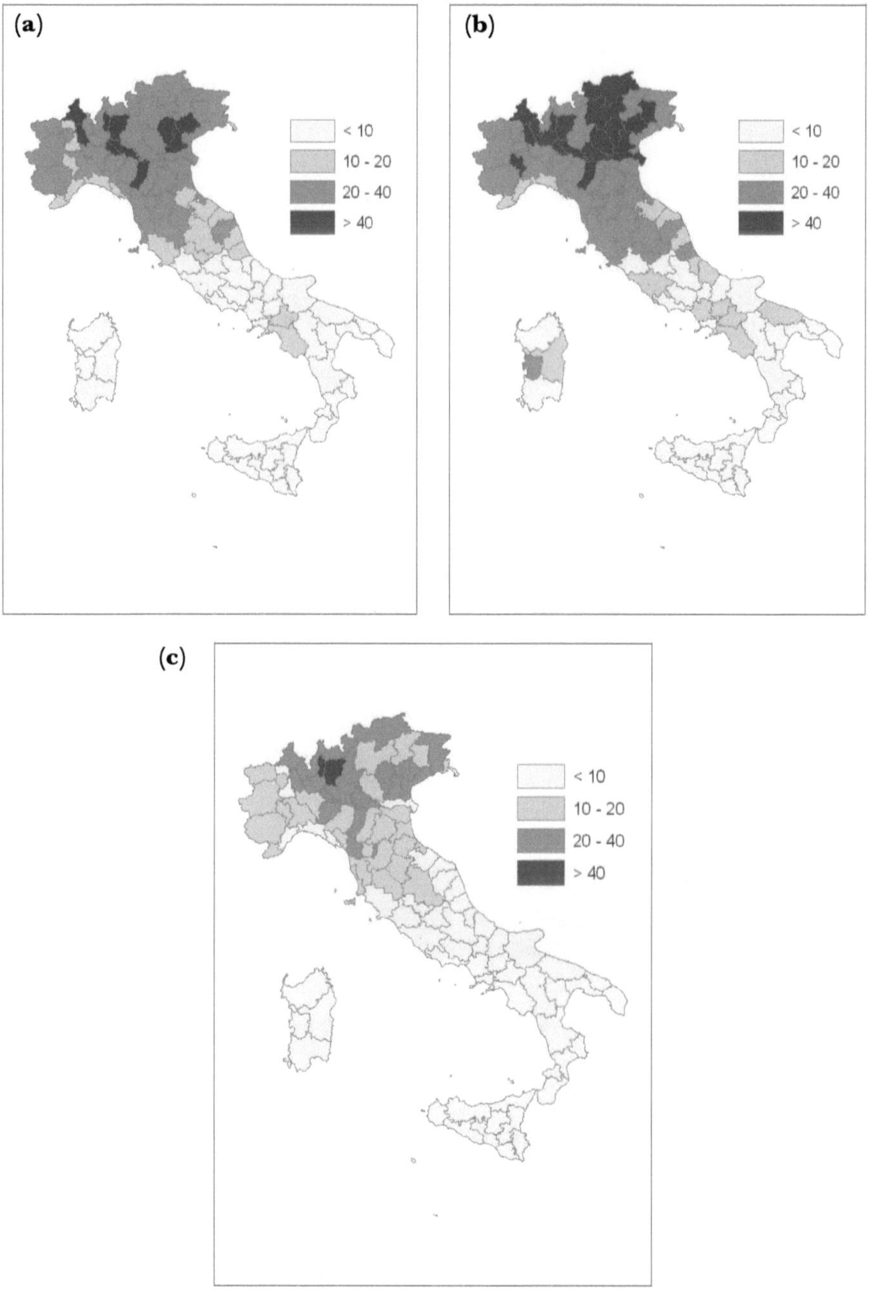

Figure 1. Tons of municipal waste subject to separate collection per 100 tons of municipal waste produced (a: 1999; b: 2002; c: 2005).

Source: Authors' elaboration of official statistics.

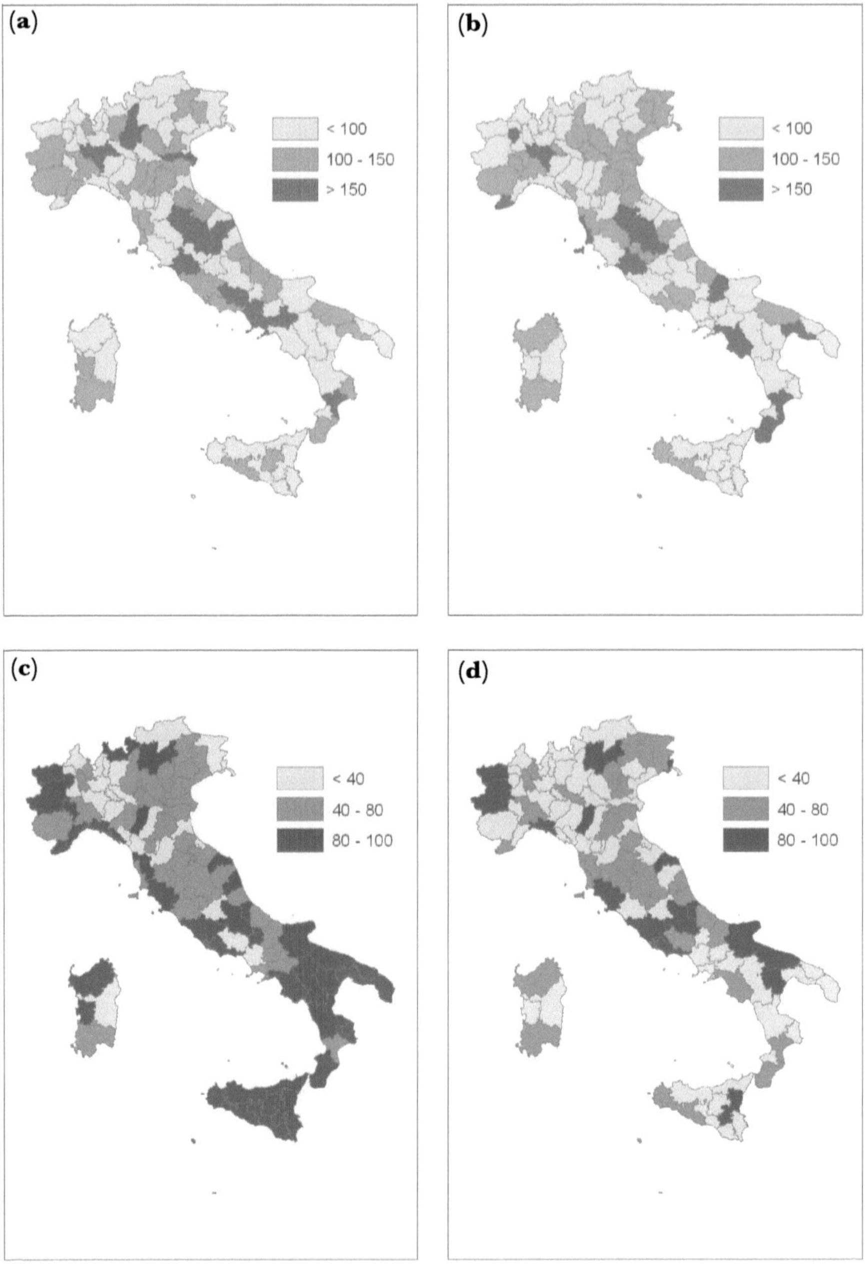

Figure 2. Top: Tons of municipal waste disposed of per 100 tons of municipal waste produced (a: 2002; b: 2005); bottom: Tons of MSW disposed of in landfill per 100 tons of MSW disposed of (c: 2002; d: 2005).

Source: Authors' elaboration of official statistics.

3.2. *Regional disparities in waste infrastructure*

In stark contrast to the South, northern regions were embarking on a significant infrastructural transition, a key enabler of the circular economy (European Commission 2020). They were actively investing in a diverse portfolio of waste treatment facilities, including composting plants, bio-stabilization units, and waste-to-energy (WtE) incinerators (Mazzanti et al. 2006). Figure 3 shows that by 2005, the density of municipal waste disposal facilities was highly uneven across the country, with a clear concentration in the North. While there was a general reduction in the number of landfills between 1999 and 2006 (Figure 4), the South's infrastructural deficit in modern treatment options meant that landfilling remained the only viable, albeit unsustainable, option for the majority of its waste.

The number of modern treatment plants was growing everywhere, but from vastly different starting points. Figure 5 highlights the situation in 2001, where the provision of treatment plants (dark purple) was heavily concentrated in the North, while most southern regions (light purple) had a provision significantly lower than the national average. Between 2000 and 2005, the number of such plants grew by about 70% in the North-West and 10% in the North-East.

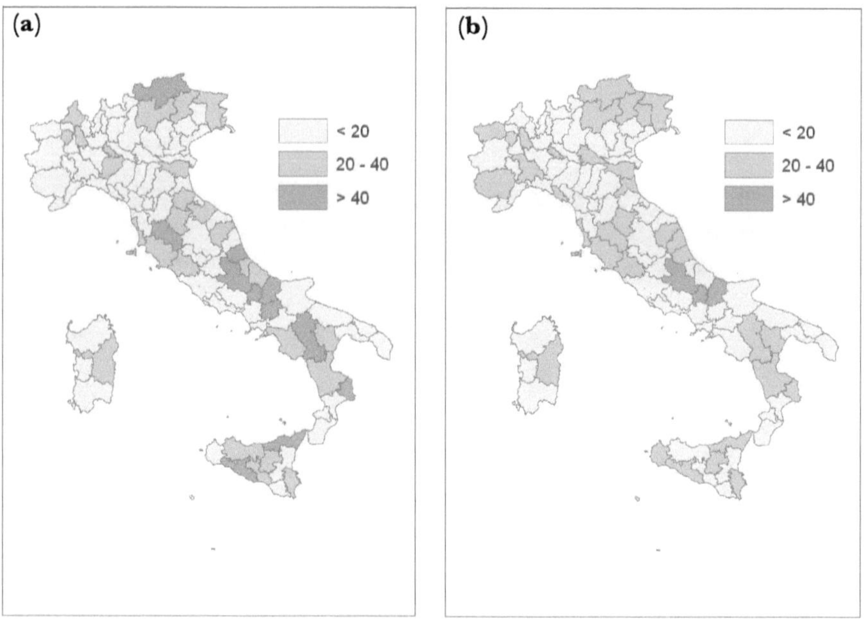

Figure 3. Municipal waste disposal facilities per 1,000,000 inhabitants (a: 2001; b: 2005).

Source: Authors' elaboration of official statistics.

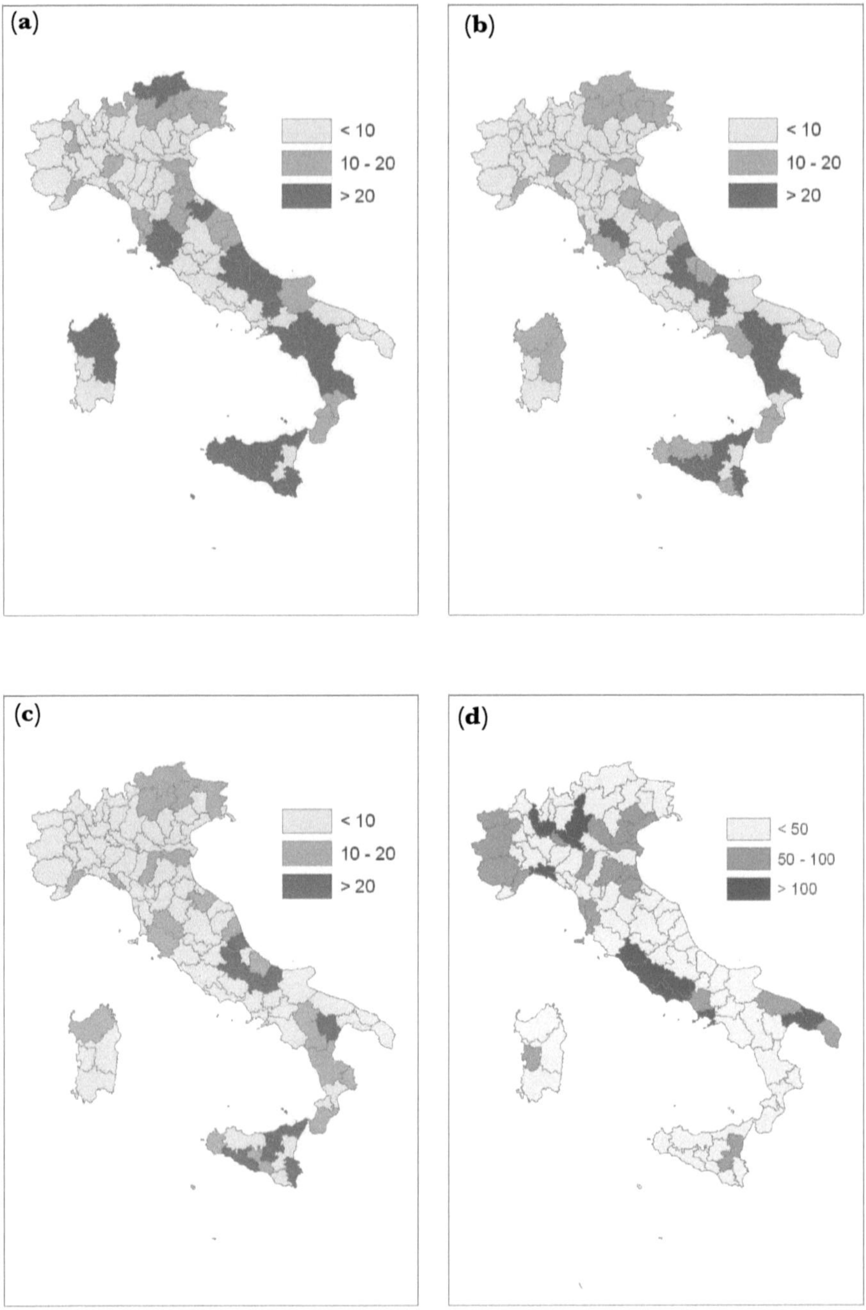

Figure 3. (*Continued*)

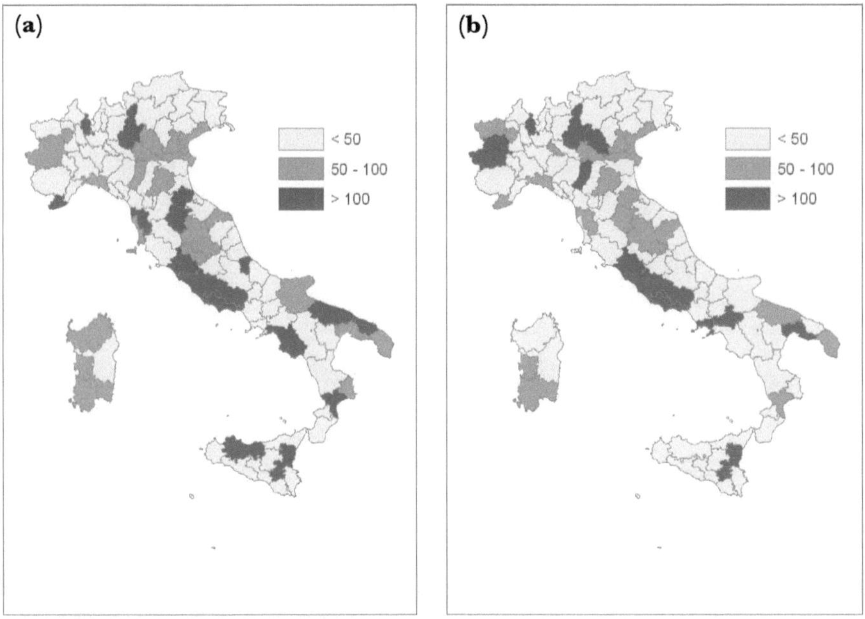

Figure 4. Top: Landfill facilities for municipal waste per 1,000,000 inhabitants (a, 1999; b: 2002; c: 2005); bottom: Average amount of municipal waste disposed of in landfills in thousands of tonnes (d: 1999; e: 2002; f: 2005).

Source: Authors' elaboration of official statistics.

While the South saw a fourfold increase, this was from an extremely low baseline, leaving its overall provision of modern infrastructure far behind (Agovino et al. 2016). This infrastructural gap is further emphasized by the capacity of incineration plants, a key technology for energy recovery from non-recyclable waste (Murphy et al. 2025; Cernuschi et al. 2021). Figures 6 and 7 clearly show that between 2002 and 2005, both the absolute and relative capacity of incinerators were heavily concentrated in the northern regions. This created a powerful path dependency, locking the regions into different environmental and economic trajectories and cementing a pattern of spatial injustice (Cerqueti et al. 2021). The debate over incineration was fierce, particularly in Campania, where the eventual construction of the Acerra incinerator was seen by the government as the end of the crisis, but was met with intense local opposition due to health and environmental concerns (Corona and Fortini 2012; D'Alisa et al. 2010).

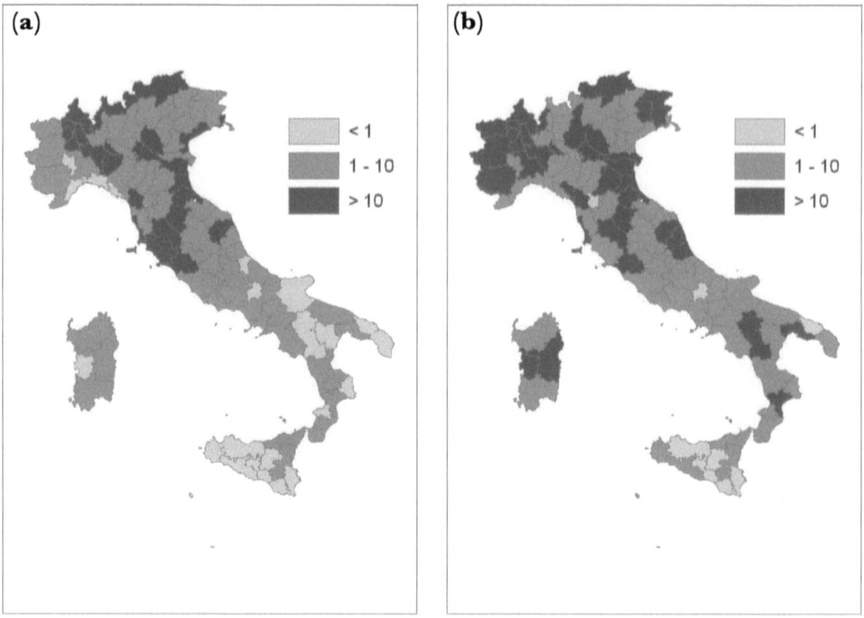

Figure 5. MSW treatment plants (composting, biostabilization and incineration) per 1,000,000 inhabitants (a: 2001; b: 2005).

Source: Authors' elaboration of official statistics.

3.3. *Challenges in special waste management*

Beyond municipal waste, the management of special (industrial and hazardous) waste also revealed significant territorial imbalances and sinister undertones (ASviS 2018). These waste streams, regulated under dedicated codes like the European Waste Catalogue, require specialized treatment and disposal infrastructure to prevent harm to human health and the environment (SDG 3.9) (WM 2025; Ambiente Diritto 2006). As shown in Figure 8, the provision of landfill facilities for special waste, measured per 10,000 km^2 of land surface, was notably higher and more consistent in the North between 1999 and 2004. This disparity was linked to the higher level of industrial development in the northern regions, which created a greater need for such facilities (Mazzanti et al. 2006).

However, this seemingly rational distribution masks a darker reality: the rise of the 'Ecomafia' (Legambiente 2024a). Organized crime syndicates, particularly the Camorra in Campania, exploited the regulatory and infrastructural gaps to create a highly lucrative business in the illegal trafficking and disposal of industrial and toxic waste, often transporting it from the industrialized North to

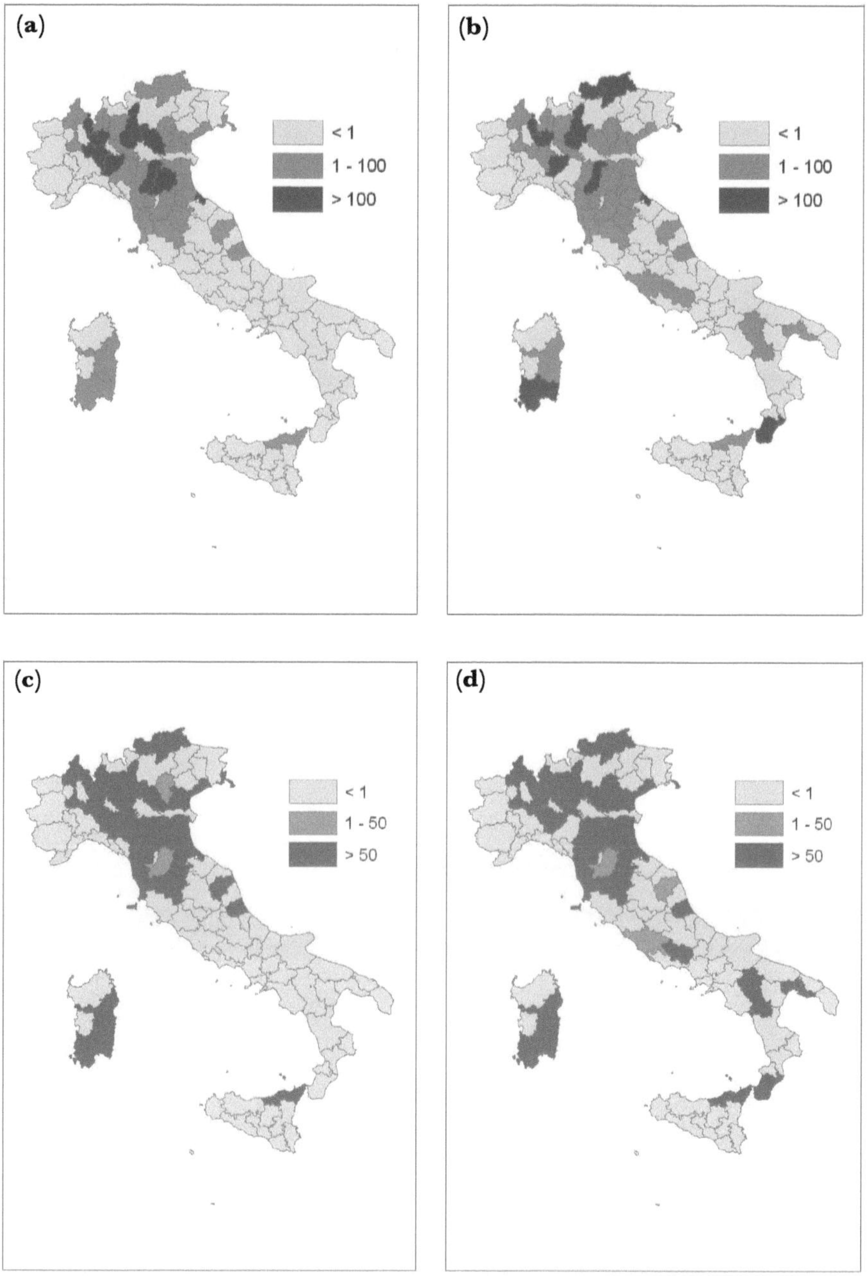

Figure 6. Top: Average annual capacity of MSW incineration plants in thousands of tonnes (a: 2002; b: 2005); bottom: Average annual capacity of MSW incineration plants per 1,000 inhabitants (c, 2002; d: 2005).

Source: Authors' elaboration of official statistics.

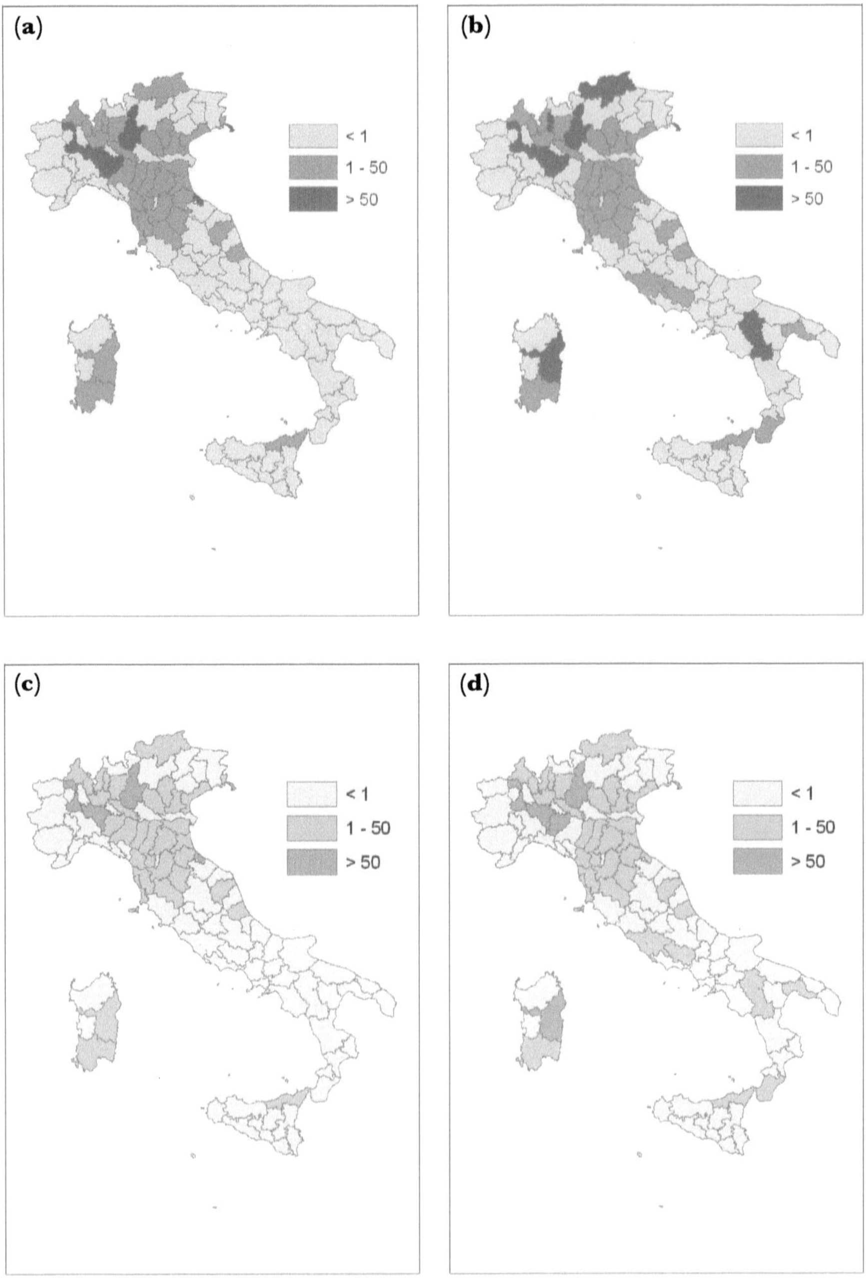

Figure 7. Top: Average annual capacity of MSW incineration plants per 100 tonnes of MSW produced (a: 2002; b: 2005); bottom: MSW treated in incineration plants per 100 tonnes of MSW produced (c: 2002; d: 2005).

Source: Authors' elaboration of official statistics.

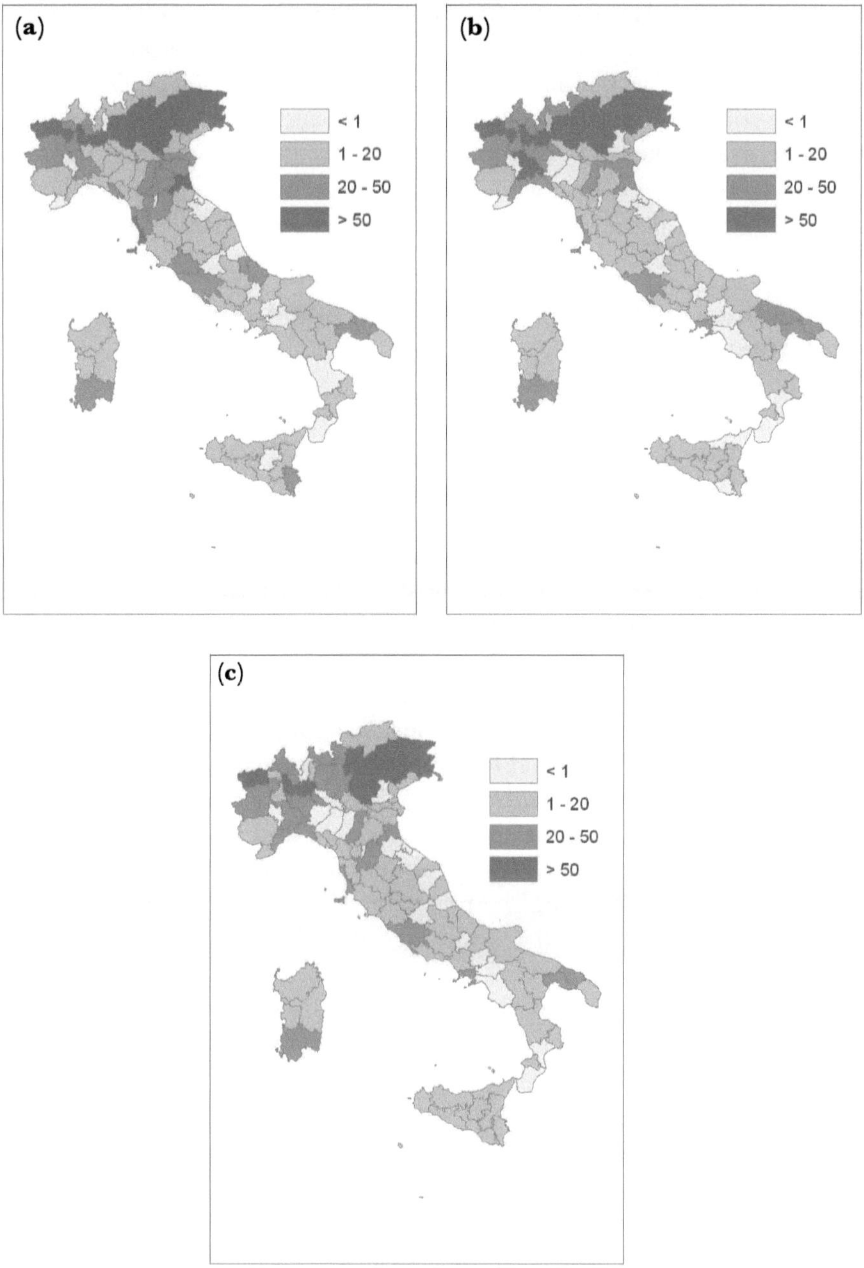

Figure 8. Landfill facilities for special waste per 10,000 km^2 of land surface (a: 1999; b: 2001; c: 2004).

Source: Authors' elaboration of official statistics.

be illegally dumped or burned in the South (D'Alisa et al. 2010). This criminal enterprise, valued in the billions of euros, involved mixing hazardous materials with municipal waste, burying it in illegal dumps and setting it on fire, leading to severe soil, water and air pollution (Legambiente 2024b). This illegal flow of waste exacerbated the environmental burden on southern communities, turning a problem of infrastructural deficit into one of profound environmental injustice and organized criminality (Senior and Mazza 2004).

4. Functional Indicators and Multivariate Analysis of Waste Dynamics

4.1. *Defining functional indicators*

To move beyond anecdotal observations and systematically investigate the territorial disparities in waste management, it is essential to define and utilize functional indicators that capture the multifaceted nature of system performance (Lo Storto 2021). Key indicators, as highlighted in the original analysis, include the share of separate waste collection relative to the total waste produced and the average quantity of municipal waste disposed of in landfills (Agovino et al. 2016). European regulations have long promoted separate collection as the principal method for managing MSW, as it is a prerequisite for high-quality recycling and for moving materials up the waste hierarchy, away from disposal (Paleari and Ferraris 2013; European Commission 2020).

The data from the pre-crisis period clearly show that while all Italian regions were adapting to these directives, they did so at vastly different speeds (Di Foggia and Beccarello 2021). Northern regions achieved the highest values in separate collection, Central regions showed gradual improvement, while Southern regions and the Islands lagged significantly, where it remained a 'not very widespread practice' (Agovino et al. 2016). This heterogeneity has been linked to differences in economic and social development, which translates into diverse capacities to plan, monitor and operate efficient waste management systems (Cerqueti et al. 2021; Colivicchi et al. 2025). Furthermore, the interpretation of landfill data requires caution. As noted in the original study, a simple count of facilities or total tonnage can be misleading without data on inter-regional waste flows. A region might have few but very large landfills, or it might be importing waste from other areas, masking the true efficiency of its local system (D'Amato et al. 2015). Socio-economic factors such as human capital and social capital also play a crucial role, as more educated and civically engaged populations tend to exhibit higher pro-environmental behaviours, including participation in recycling schemes (Agovino et al. 2017).

4.2. *Methodology: exploratory multivariate statistical analysis*

To systematically investigate these complex and interrelated territorial disparities, an exploratory multivariate statistical analysis was employed. This approach, specifically a PCA, is designed to distil the complex, multidimensional nature of waste management performance into a clearer, more interpretable structure (Jolliffe 2002). The analysis was conducted on a dataset comprising 103 Italian provinces, using data from the early 2000s sourced from official statistics (ISTAT 2019). A set of ten key descriptors was selected to create a holistic profile of each province, avoiding the limitations of single-metric evaluations (Gastaldi et al. 2020):

- Total volume of urban waste produced (kg per inhabitant);
- Percentage of waste collected via separate collection;
- Volume of specific materials collected separately (paper, glass, plastic, metal, organic, and others, in kg per inhabitant);
- Volume of selective separate collection (kg per inhabitant);
- Percentage of the population covered by the separate collection service.

The core logic of PCA is to reduce the complexity of these ten correlated variables into a smaller number of uncorrelated, latent dimensions or 'factors' (Abdi and Williams 2010). These factors represent the underlying drivers of performance that explain the maximum amount of variance in the original dataset. This provides a robust, data-driven method for identifying and visualizing the structural patterns of waste governance across Italy, a technique validated by numerous studies on waste management efficiency (Romano and Molinos-Senante 2020; lo Storto 2021). The analysis matrix was standardized to ensure comparability across variables, and from this, factor loadings (correlations between variables and factors) and factor scores (the position of each province along the new factors) were calculated to profile the territories (Agovino et al. 2016).

4.3. *Empirical results: unpacking waste management efficiency and spatial divides*

The PCA successfully extracted three significant latent factors, which together explained 63.4% of the total variance in the provincial data. The first two factors were the most powerful, accounting for 41% and 12.3% of the variance, respectively, and providing a clear lens through which to interpret the landscape of Italian waste management (Agovino et al. 2016).

Factor 1: The Efficiency and Modernization Axis. The first and most dominant factor can be unequivocally interpreted as an index of system efficiency and modernization. As shown in Table 1, variables with the highest positive correlation (factor loading) are the overall percentage of separate waste collection (0.870) and the per-capita collection of key recyclable materials like paper (0.803), glass (0.787) and organic material (0.785). A high score on this axis thus signifies a province with a well-developed, effective system for diverting waste from landfills and capturing materials for recovery – the cornerstone of a modern, circular-ready waste management strategy (Ellen MacArthur Foundation n.d.).

Factor 2: The Urban Pressure Axis. The second factor isolates a distinct dimension: urban pressure. This axis is strongly and positively correlated with the per-capita volume of urban waste produced (0.707). This highlights that, separate from management efficiency, the sheer scale of consumption and waste generation in major metropolitan areas – often linked to higher economic activity and population density – presents a unique and significant strain on infrastructure and logistics (Mazzanti et al. 2006).

Figure 9 provides a clear visualization of these results. The Cartesian diagram plots the provinces based on their scores for the first two factors. The first axis (horizontal) reveals a distinct North–South gradient: northern provinces (blue diamonds) are clustered on the right side, indicating high efficiency, while southern provinces (yellow triangles) are concentrated on

Table 1. Positioning of the analysis variables on the factorial plane (axes 1 and 2).

Variable	Axis 1	Axis 2
Urban (kg/in)	0.389	0.707
Differentiated (%)	0.870	−0.280
Paper (kg/in)	0.803	0.034
Glass (kg/in)	0.787	−0.413
Plastic (kg/in)	0.749	−0.177
Metals (kg/in)	0.495	0.181
Selective (kg/in)	0.490	0.417
Organic (kg/in)	0.785	−0.044
Other (kg/in)	0.298	0.486
Pop_covered (%)	0.424	−0.012

Source: Authors' elaboration of official statistics.

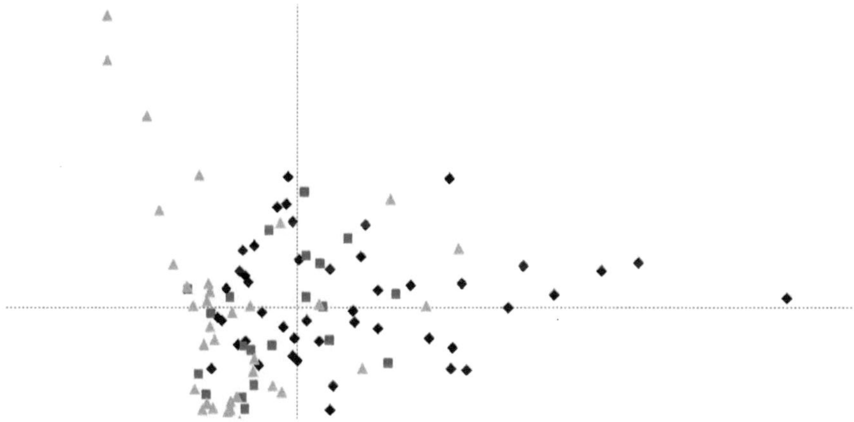

Figure 9. Positioning of the Italian provinces, distinguished by geographical distribution (triangle: Northern regions; square: Central regions; rhombus: Southern regions), on the factorial plane formed by the first two axes (x-axis, Component 1, horizontal; y-axis, Component 2, vertical).

Source: authors' elaboration of official statistics.

the left, indicating low efficiency. The second axis (vertical) captures the urban dimension, with larger metropolitan centres tending to appear in the upper part of the graph. This empirically confirms the existence of two separate, self-reinforcing systems: a virtuous cycle in the North, where investment and public participation drive efficiency and a vicious cycle in the South, where underinvestment and institutional weakness perpetuate poor performance (Ajzen 1991; Cerqueti et al. 2021). The high performance in the North can be partly explained by the Theory of Planned Behaviour, where reliable services create high perceived behavioural control, which, combined with positive social norms, encourages high citizen participation (Ajzen 1991).

By mapping the efficiency indicator derived from the first factor, Figure 10 further confirms these findings. The map distinguishes provinces into three classes of increasing efficiency (green for high, yellow for medium, red for low). It shows that the key to achieving excellent performance in separate waste collection is a combination of geographical location (in the North) and urban size (medium-small cities). The best scores (green) are found almost exclusively in northern areas with a high QoL and greater environmental awareness, while the divide between North and South remains stark, with the South dominated by low-performing (red) provinces. This map serves as a powerful indictment of the failure to achieve equitable development and environmental protection

Figure 10. Waste management and territorial infrastructure: the positioning of Italian provinces through multivariate analysis.

Source: Authors' elaboration of official statistics.

across the national territory, a failure rooted in deep-seated differences in institutional quality, governance and socio-economic context (OECD 2013; ASviS 2021; Di Foggia and Beccarello 2021).

5. Concluding Remarks

In the early 2000s, Italy's waste management system exposed profound spatial disparities that mirrored broader socio-economic divides between its northern and southern regions (Cerqueti et al. 2021). While the North saw gradual improvements in collection, recycling and treatment infrastructure, the South lagged, relying on outdated systems and landfilling (Agovino et al. 2016). This uneven progress underscored a trajectory of unsustainable urban development, incompatible with rising populations and consumption (Mazzanti et al. 2006). Inefficient waste management not only strained local environments but also eroded public trust and exacerbated regional inequalities, making it clear that a development path characterized by weak governance and reactive strategies was untenable (OECD 2013).

The analysis confirms that the infrastructure we build – or fail to build – is a physical manifestation of our social and political priorities. The North-South divide is not a temporary lag but a deeply embedded spatial injustice, the result of uneven investment and institutional fragility (Di Foggia and Beccarello 2021). A transition to a circular economy is therefore not merely a technical challenge but a fundamental socio-political project (UN-Habitat 2023). A circular economy that is not also a just circular economy is ultimately unsustainable (Aasavari 2023). If the transition replicates existing patterns of inequality – where the benefits of circularity accrue to the affluent and the burdens of residual waste fall on the marginalized – it will fail to gain social legitimacy and betray the core promise of sustainable development (Eco Cycle 2022; Circular Economy Alliance 2023). This requires a focus on distributive justice (fair sharing of costs and benefits), procedural justice (inclusive decision-making) and recognitional justice (respecting the needs of all communities) (Aasavari 2023).

Looking ahead, future research and policy must focus on a multi-pronged strategy to foster a truly national and equitable circular transition. First, there is a need for infrastructure as reparation: a national plan that explicitly targets the South not merely to close a quantitative gap, but to catalyse a qualitative leap with investment in advanced material recovery, composting, and remanufacturing facilities. Second, empowering local governance is critical.

The vicious cycle of failure in the South is fuelled by weak institutional capacity. Robust, sustained programs – such as ENEA's Arcadia Project, which builds life-cycle governance capacities among public institutions – are needed to move beyond crisis response towards strategic environmental planning (Colivicchi et al. 2025; ENEA 2023).Third, fostering a just transition through social dialogue is paramount. To rebuild trust, future planning must be rooted in transparent and inclusive stakeholder consultation, giving communities a genuine seat at the table to ensure projects meet local needs and achieve a social license to operate (UN-Habitat 2023; Ellis and Lloyd 2024). Finally, data for democratic accountability must be strengthened. The continued collection and transparent dissemination of high-quality, disaggregated data are the bedrock of accountability, allowing citizens and civil society to track progress and hold institutions responsible for ensuring the benefits of the circular economy are distributed equitably across the entire national territory (ASviS 2021; SDG Knowledge Hub 2024). Addressing these themes is essential not only for resolving Italy's enduring waste disparities but also for laying the groundwork for a more balanced and sustainable national development strategy.

References

Abdi, Hervé, and Lynne J. Williams. 2010. "Principal Component Analysis." *Wiley Interdisciplinary Reviews: Computational Statistics* 2 (4): 433–459. https://doi.org/10.1002/wics.101.

Agovino, Massimiliano, Maria Ferrara, and Antonio Garofalo. 2016. An Exploratory Analysis on Waste Management in Italy: A Focus on Waste Disposed in Landfill. *Land Use Policy* 57: 669–681. DOI: 10.1016/j.landusepol.2016.06.027.

Agovino, Massimiliano, Maria Ferrara, and Antonio Garofalo. 2017. The Driving Factors of Separate Waste Collection in Italy: A Multidimensional Analysis at Provincial Level. *Environment, Development and Sustainability* 19 (6): 2297–2316. doi: 10.1007/s10668-016-9857-9.

Ajzen, Icek. 1991. "The Theory of Planned Behaviour." *Organizational Behavior and Human Decision Processes* 50 (2): 179–211. https://doi.org/10.1016/0749-5978(91)90020-T.

Ambiente Diritto. 2006. "Codice dell'Ambiente: Decreto Legislativo 3 aprile 2006, n. 152." Gazzetta Ufficiale, n. 88, Suppl. Ordinario n. 96, April 14, 2006. Accessed July 01, 2025. https://www.ambientediritto.it/legislazione/codice-dellambiente/.

ASviS (Italian Alliance for Sustainable Development). 2018. *L'Italia e gli Obiettivi di Sviluppo Sostenibile.Rapporto ASviS* 2018. Rome: ASviS. https://asvis.it/public/asvis/files/ASviS_REPORT_2018_Ristampa.pdf.

ASviS (Italian Alliance for Sustainable Development). 2020. *Italy and the Sustainable Development Goals: ASviS Report 2020*. Rome: ASviS. https://asvis.it/public/asvis2/files/Rapporto_ASviS/Rapporto_ASviS_2020/Report_ASviS_2020_ENG_final.pdf.

ASviS (Italian Alliance for Sustainable Development). 2021. *Italy and the Sustainable Development Goals: ASviS Report 2021*. Rome: ASviS. https://asvis.it/public/asvis2/files/Rapporto_ASviS/Rapporto_2021/Report_ASviS_ENG_2021_1_.pdf.

ASviS (Italian Alliance for Sustainable Development). 2023. *Italy and the Sustainable Development Goals: ASviS Report 2023*. Rome: ASviS. https://asvis.it/public/asvis2/files/Rapporto_ASviS/Rapporto_ASViS_2023/Report_ASviS_ENG_2023_FINAL.pdf.

ASviS (Italian Alliance for Sustainable Development). 2024. *Coltivare ora il nostro futuro: Rapporto ASviS 2024.* Rome: ASviS. https://asvis.it/public/asvis2/files/Rapporto_ASviS/Rapporto_ASviS_2024/Rapporto_ASviS_2024.pdf.

Cernuschi Stefano, Mario Grosso, Federico Viganò, Maria Chiara Zanetti, Deborah Panepinto, Marco Ragazzi, Andrea Magrini, and Francesco Lombardi. 2021. *White Paper on Municipal Waste Incineration.* CEWEP–Utilitalia. https://www.cewep.eu/wp-content/uploads/2021/03/WHITE-PAPER-DEFINITIVO-2-24-febbraio-2021.pdf.

Cerqueti, Roy, Matteo Cinelli, and Leo Fulvio Minervini. 2021. "Municipal Waste Management: A Complex Network Approach with an Application to Italy." *Waste Management* 126: 597–607. https://doi.org/10.1016/j.wasman.2021.03.035.

Circular Economy Alliance. 2023. "Promoting Social Justice Through Circular Economy: Opportunities for Advocates." *Circular Economy Alliance (blog), September 26, 2023.* https://circulareconomyalliance.com/cea-blogs/promoting-social-justice-through-circular-economy-opportunities-for-advocates/.

Circular Economy Network. 2024. *Rapporto sull'economia circolare in Italia 2024: Indagine sull'economia circolare nelle piccole imprese.* Fondazione per lo Sviluppo Sostenibile. Roma, Italy.

Colivicchi, Ilaria, Pietro Gasparri, and Ginevra V. Lombardi. 2025. "Analyzing Municipal Solid Waste Management in Italy: A Machine Learning Approach to Identifying Key Drivers." *SSRN.* https://ssrn.com/abstract=5283171.

Corona, Gabriella, and Daniele Fortini. 2012. *The Problem of Waste Disposal in a Large European City: Garbage in Naples.* Lewiston, NY: Edwin Mellen Press.

D'Alisa, Giacomo, David Burgalassi, Hali Healy, and Mariana Walter. 2010. "Conflict in Campania: Waste Emergency or Crisis of Democracy." *Ecological Economics* 70 (2): 239–249. https://doi.org/10.1016/j.ecolecon.2010.06.021.

D'Amato, Alessio, Massimiliano Mazzanti, and Francesco Nicolli. 2015. "Waste and Organized Crime in Regional Environments." *Resource and Energy Economics* 41: 185–201. https://doi.org/10.1016/j.reseneeco.2015.04.003

Di Foggia, Giacomo, and Massimo Beccarello. 2021. "Designing waste management systems to meet circular economy goals: The Italian case." *Sustainable Production and Consumption* 26: 1074–1083. https://doi.org/10.1016/j.spc.2021.01.002.

Domorenok, Ekaterina. 2018. *Governing Sustainability in the EU: From Political Discourse to Policy Practices.* London: Routledge. https://doi.org/10.4324/9781315315003.

Eco Cycle. 2022. "Zero Waste and Social Justice." Ecocycle. Accessed July 3, 2025. https://ecocycle.org/learn-about-zero-waste/zero-waste-and-social-justice/.

Ellen MacArthur Foundation. n.d. "Circular economy introduction." Ellen MacArthur Foundation. Accessed July 3, 2025. https://www.ellenmacarthurfoundation.org/topics/circular-economy-introduction/overview.

ENEA (Italian National Agency for New Technologies, Energy and Sustainable Economic Development). 2023. "*Progetto Arcadia: Approccio ciclo di vita nei contratti pubblici e banca dati italiana LCA per l'uso efficiente delle risorse.*" ENEA. https://sostenibilita.enea.it/projects/arcadia.

European Commission. 2020. *Circular economy action plan – For a cleaner and more competitive Europe*, Publications Office of the European Union. https://data.europa.eu/doi/10.2779/05068.

Gastaldi, Massimo, Ginevra Virginia Lombardi, Agnese Rapposelli, and Giulia Romano. 2020. "The Efficiency of Waste Sector in Italy: An Application by Data Envelopment

Analysis." *Environmental and Climate Technologies*, 24 (3): 225–238. doi:10.2478/rtuect-2020-0099.

GIZ (Deutsche Gesellschaft für Internationale Zusammenarbeit). 2024. *Towards a Sustainable Future: Recommendations for a Just Transition in Waste Management and Circular Economy in the ASEAN Region*. Deutsche Gesellschaft für Internationale Zusammenarbeit (GIZ) GmbH, Bonn, Germany. https://www.giz.de/en/downloads/giz2024-en-just-transition-report.pdf.

Ellis, Michael, and Colleen Lloyd, 2024. "*Just Transition: Blending Social and Environmental Considerations*." ISS Governance/Insights. Accessed June 26, 2025. https://www.issgovernance.com/library/just-transition-blending-social-and-environmental-considerations/

ILO (International Labour Organization). 2015. *Guidelines for a Just Transition towards Environmentally Sustainable Economies and Societies for All*. Geneva: International Labour Organization.https://www.ilo.org/sites/default/files/wcmsp5/groups/public/%40ed_emp/%40emp_ent/documents/publication/wcms_432859.pdf.

ISTAT (Istituto Nazionale di Statistica). 2019. *SDGs Report 2019: Statistical Information for 2030 Agenda in Italy*. ISTAT: Rome. https://www.istat.it/wp-content/uploads/2019/08/SDGs-2019_inglese.pdf.

ISPRA (Istituto Superiore per la Protezione e la Ricerca Ambientale). 2024. *Rapporto Rifiuti Urbani Edizione 2024*. ISPRA: Rome.

IUCN (International Union for Conservation of Nature). 2024. "Just transition and inclusion: overcoming socio-economic barriers and driving a circular transition." IUCN, accessed June 8, 2025. https://iucn.org/story/202411/just-transition-and-inclusion-overcoming-socio-economic-barriers-and-driving-circular.

Jolliffe, Ian T. 2002. *Principal Component Analysis*. 2nd ed. New York: Springer.

Joshi, Aasavari. 2023. "Ensuring a Fair and Just Transition to a Circular Economy." Circular Innovation Lab, accessed July 22, 2025. https://www.circularinnovationlab.com/post/ensuring-a-fair-and-just-transition-to-a-circular-economy.

Kaza, Silpa, Lisa C. Yao, Perinaz Bhada-Tata, and Frank Van Woerden. 2018. What a Waste 2.0: A Global Snapshot of Solid Waste Management to 2050. Urban *Development*, World Bank. http://hdl.handle.net/10986/30317.

Legambiente. 2024a. "ControEcomafie, per difenderci dall'illegalità ambientale." Legambiente. Accessed June 19, 2025. https://www.legambiente.it/news-storie/controecomafie/.

———. 2024b. *Rapporto Ecomafia 2024: Le storie e i numeri della criminalità ambientale*. Edizioni Ambiente: Rome.

Lo Storto, Corrado. 2021. Eco-Productivity Analysis of the Municipal Solid Waste Service in the Apulia Region from 2010 to 2017. *Sustainability* 13(21): 12008. https://doi.org/10.3390/su132112008.

MASE (Italian Ministry of Environment and Energy Security). 1993. *Piano Nazionale per lo Sviluppo Sostenibile*: In Attuazione Dell'Agenda 21. Rome. https://www.mase.gov.it/portale/documents/d/guest/piano_nazionale_svs_italia-pdf.

MASE (Italian Ministry of Environment and Energy Security). 2008. *Piano d'azione per la sostenibilità ambientale dei consumi nel settore della pubblica amministrazione*. Decreto Interministeriale, n. 135 e Allegato, Rome. https://www.mase.gov.it/portale/-/decreto-interministeriale-11-aprile-2008-n.-135-e-allegato-piano-d-039-azione-per-la-sostenibilita-ambientale-dei-consumi-nel-settore-della-pubblica-amministrazione-g.u.-8-maggio-2008-n.107-.

MASE (Italian Ministry of Environment and Energy Security). 2017. *Strategia Nazionale per lo Sviluppo Sostenibile*. Rapporti con l'Unione Europea e gli Organismi internazionali. Rome. https://www.regione.piemonte.it/web/media/17846/download.

Mazzanti, Massimiliano, Anna Montini, and Roberto Zoboli. 2006. "*Municipal Waste Production, Economic Drivers and New Waste Policies: EKC Evidence from Italian Regional and Provincial Panel Data*." FEEM Working Paper No. 20.2006. Milan: Fondazione Eni Enrico Mattei.

Murphy, Fionnuala, Egle Gusciute, Maneesh Kumar Mediboyina, and Antonio Calvi. 2025. *Social and Environmental Sustainability of Municipal Solid Waste in the Context of the UN Sustainable Development Goals*. IEA Bioenergy, Task 36 Report. https://www.ieabioenergy.com/wp-content/uploads/2025/03/IEA-Bioenergy-Task-36_-Social-and-Environmental-Sustainability-of-Municipal-Solid-Waste-in-the-Context-of-the-UN-Sustainable-Development-Goals_final-1.pdf.

Newton, John. 2022. "The Effects Of Landfills On The Environment." sciencing.com. Accessed June 8, 2025. https://www.sciencing.com/effects-landfills-environment-8662463/

OECD (Organisation for Economic Co-operation and Development). 2013. OECD Environmental Performance Reviews. OECD Publishing, Paris. https://doi.org/10.1787/9789264186378-en.

OECD (Organisation for Economic Co-operation and Development). 2022. *Italy's National Action Plan for Policy Coherence for Sustainable Development*. OECD Publishing, Paris. https://doi.org/10.1787/54226722-en.

Paleari, Susanna and Matteo Ferraris. 2013. *Municipal Waste Management in Italy*. European Environment Agency: Copenhagen, Denmark.

Park, Chun Kyoo, Sara C. de Hallgren, and Emily Carroll. 2024. "Closing the Loop on Waste." SDG Action. United Nations Office for Sustainable Development. Accessed June 06, 2025. https://sdg-action.org/closing-the-loop-on-waste/.

Patelli, and Daniela Ducci. 2023. "Geochemical and Isotopic Evidence for Investigating the Impacts of Landfills on Groundwater: A Case Study in the Campania Region (Southern Italy)." *Sustainability* 15(22): 15822. https://doi.org/10.3390/su152215822.

Prioreschi, Roberto, Pierluigi Serlenga, Carlo Farina, Matteo Capellini, Roberta Berlinghieri, and Ghizlene Azira. 2023. *Italy Is Falling Short of a 1.5-Degree World*. Bain & Company Report, October 25, 2023. https://www.bain.com/insights/italy-is-falling-short-of-a-1.5-degree-world-report/.

Ribellino, Marialuce. 2024. Italy at a crossroads for sustainability: the ninth ASviS Report highlights the urgency for change. *progetto Doable* (blog), 24 October 2024. https://progettodoable.it/en/resources/blog/italy-at-a-crossroads-for-sustainability-the-ninth-asvis-report-highlights-the-urgency-for-change.

Richiedei, Anna, and Michele Pezzagno. 2022. "Territorializing and monitoring of sustainable development goals in Italy: An overview." *Sustainability* 14(5): 3056. https://doi.org/10.3390/su14053056.

Romano, Gianpiero, and Maria Molinos-Senante. 2020. "Factors Affecting Eco-Efficiency of Municipal Waste Services in Tuscan Municipalities: An Empirical Investigation of Different Management Models." *Waste Management* 105: 384–94. doi: 10.1016/j.wasman.2020.02.028.

SDG Knowledge Hub. 2024. UNECE, OECD Issue Guidelines for Measuring Circular *Economy*. The International Institute for Sustainable Development

(IISD). Accessed June 07, 2025. https://sdg.iisd.org/news/unece-oecd-issue-guidelines-for-measuring-circular-economy/.

Senior, Kathryn, and Alfredo Mazza. 2004. Italian "Triangle of death" linked to waste crisis. *The Lancet Oncology*, 5(9):525–527. doi: 10.1016/s1470-2045(04)01561-x.

Stevenazzi, Stefania, Elena Del Gaudio, Domenico Ruggiero, Ciro D'Aniso, Anna Maria, and Stodola, Sarah. 2023. How Italy got its citizens – and me – to adopt a rigorous recycling scheme. Grist. Accessed July 21, 2025. https://grist.org/looking-forward/how-italy-got-its-citizens-and-me-to-adopt-a-rigorous-recycling-scheme/.

Triassi, Maria, Rossella Alfano, Maddalena Illario, Antonio Nardone, Oreste Caporale, and Paolo Montuori. 2015. "Environmental Pollution from Illegal Waste Disposal and Health Effects: A Review on the "Triangle of Death"" *International Journal of Environmental Research and Public Health* 12, no. 2: 1216-1236. https://doi.org/10.3390/ijerph120201216.

UN (United Nations). 1992. *Agenda 21: Programme of action for sustainable development.* New York: UN DPI. https://digitallibrary.un.org/record/170126.

UN (United Nations). 2015. *Transforming Our World: The 2030 Agenda for Sustainable Development.* New York: United Nations. https://digitallibrary.un.org/record/1654217?v=pdf.

UN (United Nations). 2023. *Financing for Sustainable Development Goals Report 2023.* New York: United Nations. https://desapublications.un.org/publications/financing-sustainable-development-report-2023.

UN (United Nations). 2025. *The Sustainable Development Goals Report 2025.* New York: UN DESA. https://unstats.un.org/sdgs/report/2025/.

UN-Habitat (United Nations Human Settlements Programme). 2023. *Just Transition in Waste Management and Circular Economy.* Thematic Brief. The UN Secretary-General's Advisory Board on Zero Waste. https://unhabitat.org/sites/default/files/2024/10/just_transition.pdf.

WCED (World Commission on Environment and Development). 1987. *Brundtland Report: Our Common Future.* New York: United Nations. https://www.brundtland.co.za/other-publication/brundtland-report-1987-our-common-future/

Wilson, David C. 2007. "Development Drivers for Solid Waste Management." *Waste Management and Research* 25 (3): 198–207. https://doi.org/10.1177/0734242X0707914.

———. 2015. Global Waste Management – The Way Forward. In *Global Waste Management Outlook by United Nations Environment Programme (UNEP) and International Solid Waste Association (ISWA)*, 267–304. https://www.unep.org/resources/report/global-waste-management-outlook.

WM (Waste Management Inc). 2025. *United Nations Sustainable Development Goals.* Houston: WM. https://sustainability.wm.com/downloads/WM_2025_SDG.pdf.

INDEX

www.ingramcontent.com/pod-product-compliance
Ingram Content Group UK Ltd.
Pitfield, Milton Keynes, MK11 3LW, UK
UKHW041831100926
8556IPUK00002B/123

9 781839 998751